Water Distribution System Monitoring

A Practical Approach for Evaluating Drinking Water Quality
Second Edition

Water Distribution System Monitoring

A Practical Approach for Evaluating Drinking Water Quality
Second Edition

Abigail F. Cantor

CRC Press
Taylor & Francis Group
Boca Raton London New York

CRC Press is an imprint of the
Taylor & Francis Group, an **informa** business

CRC Press
Taylor & Francis Group
6000 Broken Sound Parkway NW, Suite 300
Boca Raton, FL 33487-2742

International Standard Book Number-13: 978-1-1380-6403-4 (Paperback)
International Standard Book Number-13: 978-0-8153-7495-4 (Hardback)

Library of Congress Cataloging-in-Publication Data

Names: Cantor, Abigail F., author.
Title: Water distribution system monitoring / Abigail F. Cantor.
Description: 2nd edition. | Boca Raton : Taylor & Francis, CRC Press, 2017. | Includes bibliographical references.
Identifiers: LCCN 2017037273| ISBN 9781138064034 (pbk.) | ISBN 9780815374954 (hardback) | ISBN 9781315160634 (ebook)
Subjects: LCSH: Water-pipes--Monitoring. | Water quality--Measurement. | Water--Distribution--Quality control. | Drinking water--Lead content. | Drinking water--Copper content.
Classification: LCC TD491 .C363 2017 | DDC 628.1/5--dc23
LC record available at https://lccn.loc.gov/2017037273

Visit the Taylor & Francis Web site at
http://www.taylorandfrancis.com

and the CRC Press Web site at
http://www.crcpress.com

Contents

Contents

List of Figures

List of Tables

Acknowledgments

It has been nine years since the first edition of this book. This has been nine more years of working with enthusiastic people focused on improving drinking water quality. There are many people to thank for this wonderful collaboration and exchange of ideas.

But some water system personnel prefer anonymity. Just know that there are many people with inquisitive minds putting their heart and soul into providing drinking water of high quality.

Below is a list of people and organizations that I am able to bring to your attention.

A project funded by the Water Research Foundation of Denver, Colorado, the Water Environment Research Foundation of Alexandria, Virginia, and five water utilities in Wisconsin allowed me to pull together work on this monitoring approach and the comprehensive perspective of water quality that I began developing around 1995. The project report is titled *Project 4586: Optimization of Phosphorus-Based Corrosion Control Chemicals Using a Comprehensive Perspective of Water Quality* (Cantor 2017). It can be downloaded at no cost from the Water Research Foundation website, http://www.waterrf.org. Both the report and this book use materials that have been repeated in many of my publications and presentations over the years. This book starts out by describing a comprehensive perspective of water quality; the research project allowed me to formally demonstrate that more comprehensive measurement of factors in a water system lead to the identification of significant factors shaping the water quality beyond those described in drinking water regulations. Much of the information in Chapter 4, "Strategic Planning for Improved Water Quality," and Chapter 5, "Data Management and Analysis," was used in both documents. In Chapter 6, "Case Studies," Water Systems 1 and 2 were the subjects of study in the research project. In general, the techniques described in this book were utilized in the research project. The reader of this book is encouraged to download the research report to experience the monitoring and assessment techniques in action.

In addition, some of this material was repeated from my contributions to the American Water Works Association's (Denver, Colorado) *Manual of Water Supply Practices M-58: Internal Corrosion Control in Water Distribution Systems (2nd Edition)* (AWWA 2017). American Water Works Association publications can be found at their website, http://www.awwa.org. The materials that I have repeated in several publications were actually developed in teaching a professional development course at the University of Wisconsin, *Water Quality in Distribution Systems and Building Plumbing* (https://edp.wisc.edu/course/water-quality-in-distribution-systems-and-building-plumbing/). Ned Paschke, the director of the program, has a reputation for developing classes communicating the cutting edge of engineering concepts and technologies.

I do want to point out one particular water utility manager. Eric Kiefer, Manager of the North Shore Water Commission water treatment plant in Glendale, Wisconsin, is an "early adopter" (Rogers 2003) of ideas and technologies. He has involved the PRS Monitoring Station in two Water Research Foundation projects (Project 4286 and Project 4586) as well as in making a critical operational change at the water treatment plant. His involvement with the PRS Monitoring Station has called the technique to the attention of the Water Research Foundation and others in the drinking water field.

Another water utility manager who has played a major role in the development of the PRS Monitoring Station and the comprehensive monitoring technique is Nancy Quirk, P.E. In 2005, she was the Technical Services Manager for Waukesha Wisconsin Water Utility when she requested that I determine if the water system remained in compliance with the Lead and Copper Rule after a water treatment change. I told her about an idea that I had for a distribution system monitoring station and monitoring strategy. She liked the idea and the first PRS Monitoring Station was built from my design by personnel at the Waukesha Water Utility in 2006. Jumping over the years to 2013, Ms. Quirk was now General Manager of the Green Bay Wisconsin Water Utility. She requested help with Lead and Copper Rule compliance. The latest version of the PRS Monitoring Station was used to sleuth out the mechanisms causing elevated lead in the water system. You can read about the experience regarding Water System 1 in Chapter 6, "Case Studies." It is equally important to thank Russ Hardwick, Water Quality Manager of the Green Bay Water Utility, who deftly carried out the day-to-day sampling of the system.

There are others whose work has been important for the interpretation of the monitoring data obtained using my comprehensive strategy. Dr. J. Barry Maynard, currently retired geology professor from the University of Cincinnati, has analyzed the chemical scales on the PRS Monitoring Station metal plates since 2006 when the first PRS Monitoring Station was installed. His insight into chemical scales in drinking water systems has added an important dimension to the use of PRS Monitoring Stations and his analyses have uncovered key aspects of lead and copper release mechanisms.

Dr. Andrew Jacque, P.E., Founder and Chief Scientist of Water Quality Investigations, LLC in Mount Horeb, Wisconsin, has made tremendous strides in uncovering how microorganisms participate in water chemistry and in shaping water quality. I met Dr. Jacque many years ago in a mutual quest for improved methods for measuring the population of microorganisms in water and on pipe walls. We have collaborated on many projects since that time, during which time Dr. Jacque has expanded the possibilities of microbiological materials measurement and interpretations of the data.

Dr. Kevin Little, Statistician and Founder of Informing Ecological Design, LLC, has been an important mentor to me on the use of statistical techniques for the analysis of drinking water monitoring data. Dr. Little was the first to suggest that the Shewhart Control Charts be used for this purpose. He has also introduced other concepts regarding data presentation and statistical programming language. This quantity of data could not be handled without the techniques that he promotes.

Rob Spence, vice-president of the Rundle-Spence Company—plumbing, heating, well, and septic equipment distributor—and mechanical engineer, took on the task of improving the design of Process Research Solutions' Monitoring Station and assembling the stations for water utilities starting around 2009. He has made it possible to easily produce and install a monitoring station in a distribution system to get a water quality monitoring program underway.

Last, but not least, are the editors who assisted with this edition of the book. I first met Archie Degnan when he was a Senior Microbiologist at the Wisconsin State Laboratory of Hygiene in Madison. In the first edition of the book, Mr. Degnan wrote about the technique that he developed to remove the biofilms from the PRS Monitoring Station metal plates and to determine the microbiological population. Over the past eight years since the first edition, analytical techniques have changed and Mr. Degnan has retired from the laboratory. However, he works for Process Research Solutions part-time and was a technical editor on this second edition of the book.

Another editor of the second edition was Susanna Cantor. She holds a degree in communications and certification in configuration management and brings more than 40 years of writing, editing, and information management experience to her work. During her career, she has worked as a reporter, editor, administrator, and manager in the newspaper publishing industry and led a variety of system documentation and knowledge management projects at RTI International, a research institute located in Central North Carolina. Now retired, Ms. Cantor provides freelance editing support to food magazines and other publications. (If you haven't guessed, she is my sister.)

Oh yes, there is always someone who patiently waits for a writer of a book to finally finish the book and free up time. They also have to listen incessantly to topics of conversation of which they have no interest, such as corroding water pipes. In my case, that is my husband, Alan Kalker.

The development of the comprehensive approach to drinking water quality and this book would not have become a reality without this excellent team of people.

Abigail F. Cantor
Madison, Wisconsin
www.processresearch.net

Author

Abigail F. Cantor is a chemical engineer, a computer programmer, and President of Process Research Solutions, LLC in Madison, Wisconsin. She holds a B.S. in Civil/Environmental Engineering from the University of Tennessee in Knoxville and an M.S. in Chemical Engineering from Columbia University in New York City. Ms. Cantor began her professional life in 1980 as a water and wastewater treatment process design engineer working for a major engineering consulting firm.

In 1991, her focus turned toward drinking water quality investigations. This change was influenced by a newly published federal drinking water regulation concerning lead and copper, which opened up an area of water utility operation that had many knowledge gaps. The water chemistry involved intrigued her, as did the process of investigation and discovery.

In 1997, Ms. Cantor left an engineering consulting firm to work on her own. Since that time, her company, Process Research Solutions, LLC, has developed a national reputation for water quality investigations and for water quality data management computer software.

Ms. Cantor is a registered Professional Engineer in Wisconsin and is an active member of the American Water Works Association (AWWA) and its Wisconsin chapter (WI AWWA). She served for twelve years as vice-chair and chair of the WI AWWA Research Committee and has participated on the AWWA Lead and Copper Rule Task Advisory Committee critiquing the U.S. Environmental Protection Agency's communications on the Lead and Copper Rule.

The monitoring station described in this book won the 2007 WI AWWA Gimmicks and Gadgets award.

1 Introduction

The United States has been very successful in protecting drinking water quality. We do not think twice about traveling anywhere in the country and drinking water in restaurants and other buildings. However, there have been some major incidents regarding drinking water quality over recent years. A notable case involved high lead levels in Washington, DC's drinking water in 2001 (*Washington Post* 2004). Another prominent case occurred in Flint, Michigan, in 2014, where high lead, iron, disinfection by-products, as well as a presence of *E. coli* coincided with an uptick in illness from *Legionella* (Masten et al. 2016). Why were the drinking water regulations unable to prevent these catastrophes?

The answer is based on where the contaminant enters the water. Drinking water regulations have been the most effective when the contaminant enters via the source water. See Figure 1.1.

In Figure 1.1, a contaminant is found in the groundwater or surface water that is the source of the drinking water. The contaminant can be measured before it enters the water distribution system. If it is higher in concentration than limiting regulatory or voluntary health goals, then a treatment plant can be placed at the same location to remove the contaminant.

In contrast, when a contaminant enters the drinking water from within the distribution system, its concentration changes over location in the distribution system and over time. See Figure 1.2. The contaminant concentration varies like this because it is created in the distribution system from interactions in the water as it resides in the pipes and from interactions between the water and the piping material.

In this instance, instead of a discrete number of locations to study and control, as in Figure 1.1, there are an unknown number of water quality scenarios around the distribution system and many dilemmas arise:

- Where do we measure the contaminant so that it is representative of what consumers are exposed to?
- How do we measure the contaminant if every building piping system is different?
- What do we measure to determine how the contaminant is created in the distribution system?
- How are we going to prevent or treat the contaminant throughout the distribution system?

Drinking water regulations must answer these questions in order to protect consumers against each distribution-system contaminant. But water distribution systems are a dynamic complex of physical, chemical, and microbiological forces. Some drinking water regulations, such as the U.S. Environmental Protection Agency's (EPA's)

Consumers

Measure 'X' here.
Remove 'X' here.

Contaminant 'X' in
water source

Distribution system

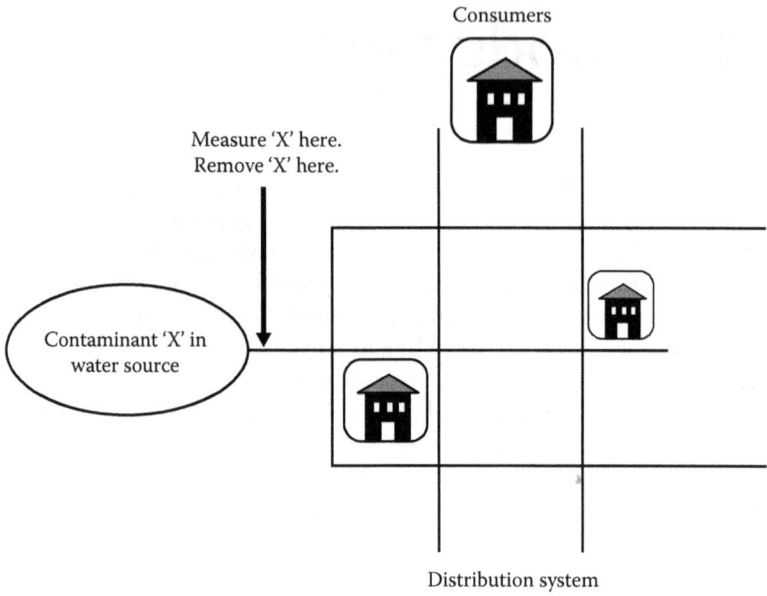

FIGURE 1.1 Contaminant in the source water.

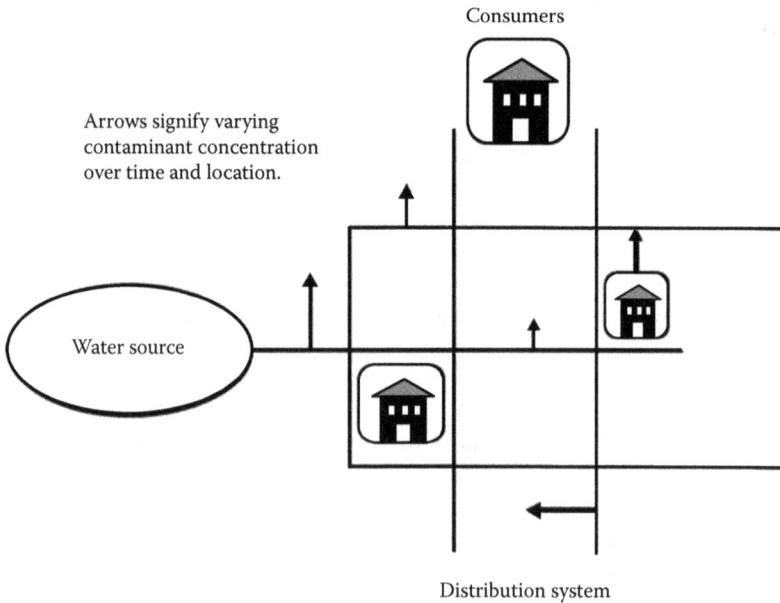

Consumers

Arrows signify varying
contaminant concentration
over time and location.

Water source

Distribution system

FIGURE 1.2 Contaminant entering the drinking water from interactions in the distribution system and varying over location and time.

Lead and Copper Rule (40 CFR Part 141 Subpart I), have not yet been able to adequately address this complexity.

This book tackles the complexity of water distribution systems. It is a practical step-by-step approach to understanding and controlling the drinking water quality that consumers experience in a water system. It demonstrates how to be proactive on water quality issues through routine monitoring of the distribution system and is written to help the practitioner prevent the devastating and costly effects of

- Falling out of compliance with the Lead and Copper Rule
- Developing pinhole leaks in water service lines and private plumbing
- Leaving the water distribution system vulnerable to microorganisms that can cause illness or can corrode metals
- Experiencing unwanted side effects from treatment chemicals
- Adding the wrong water treatment chemical or using the wrong dosage
- Alarming consumers with water that is discolored or has a bad odor
- Releasing other metals that have accumulated in the distribution system that may cause consumer health issues, such as arsenic or radium

This book describes a path to well-defined and measurable control of the distribution system water quality that consumers experience.

2 A Comprehensive Perspective of Drinking Water Quality

In order to tackle the complexity of water distribution systems, a comprehensive perspective of drinking water quality is necessary. In a comprehensive perspective, water entering distribution system piping is a complex solution of naturally occurring chemicals and microorganisms as well as added treatment chemicals.

Drinking water comes to the system from nature, where it has been in contact with soil, rocks, and air. Chemicals from those media dissolve into the water or become entrained as particulate matter. Microorganisms, which are everywhere in our environment, are also transferred to the water along with nutrients for their growth. Water treatment, when performed, does not necessarily remove all of the chemical and microbiological components in the water. Additional chemicals may also be included in the treatment.

When inside of distribution system piping—not only the water mains, but also the service lines and building plumbing—water is subjected to even more complexity as it comes in contact with pipe wall accumulations of chemical scales and biofilms that have built up over the years.

Coming out of the piping is drinking water that has been transformed by its interaction with the pipe wall accumulations. This is the process by which distribution system water quality is shaped. See Figure 2.1 and Table 2.1.

Figure 2.2 shows the many combinations of interactions that can occur inside the piping—chemistry in water, microbiology in water, chemistry in pipe wall accumulations, microbiology in pipe wall accumulations, chemistry between the water and the pipe wall accumulations, microbiological interactions between water and pipe wall accumulations, chemical and microbiological interactions in water, chemical and microbiological interactions in pipe wall accumulations, and chemical and microbiological interactions between water and pipe wall accumulations. Physical disturbances in the system, such as water velocity, pressure gradients, and vibrations, can also add to this complexity. The point is that there are no scientific formulae that can describe this complexity; the final water quality cannot be predicted.

The final water quality can include quite a number of disagreeable and harmful qualities, some of which are addressed by secondary drinking water regulations (discolored water; release of iron, manganese, and aluminum or other metals; water with odor), some of which are addressed by primary drinking water regulations (release of lead or copper, presence of pathogenic microorganisms* such as

---

* Pathogenic microorganisms refer to microorganisms that can directly cause human illness.

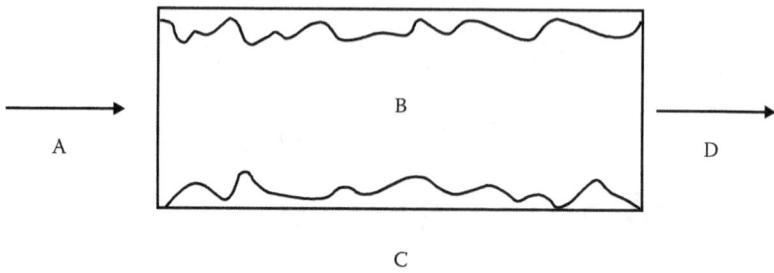

FIGURE 2.1   Two perspectives of drinking water quality.

TABLE 2.1
Two Perspectives of Drinking Water Quality from Figure 2.1

Component	Description	Comprehensive Perspective	Regulatory Perspective
A	Water flowing into pipe.	A solution of various chemical compounds (including those of organic carbon, nitrogen, phosphorus, carbonate, hydroxide, hydrogen, oxygen, carbon dioxide, sulfate, chloride, iron, manganese, other metals) and microorganisms.	A solution of carbonate compounds.
B	Interactions in pipe.	See Figure 2.2 for complex contents and interactions between chemistry and microbiology.	Carbonate compounds meet lead and copper ions released from piping material by uniform corrosion.
C	Pipe wall accumulations.	Chemical scales of a variety of solids (including carbonates, oxides, iron, manganese, aluminum, phosphorus, sulfur) and biofilms.	Insoluble fractions of lead or copper carbonate compounds.
D	Water flowing out of pipe.	All outcomes are interrelated as a result of the complex interactions in the pipe. Negative characteristics can include lead, copper, iron, and other metal release in both dissolved and particulate form, discolored water, water with odor, presence of pathogenic microorganisms such as E. Coli and Legionella, excessive growth of non-pathogenic microorganisms with potential metals corrosion, and formation of disinfection by-products.	Soluble fractions of lead or copper carbonate compounds.

Final drinking water quality

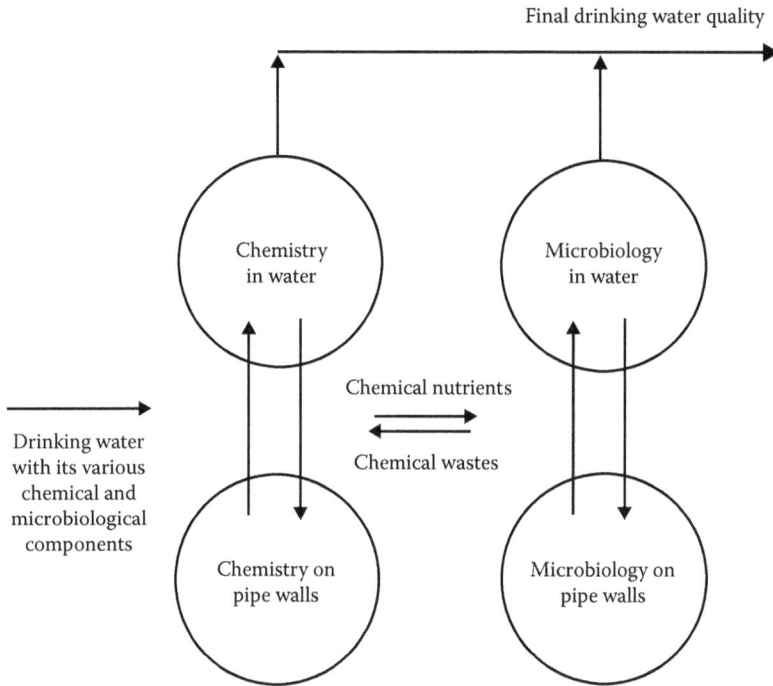

FIGURE 2.2   Complexity of water and pipe wall accumulation interactions.

E. coli, formation of disinfection by-products), and some of which are not typically addressed, such as excessive growth of nonpathogenic microorganisms or release of harmful metals, such as arsenic and radium from pipe wall accumulations. See Figure 2.1 and Table 2.1.

The point of this comprehensive perspective is that all of the water quality outcomes—both good and bad—are interrelated. They are manifestations of the same phenomena, the interactions of the complex solution of water coming in contact with the complex pipe wall accumulations of chemical scales and biofilms.

In the comprehensive perspective, the remedy for bad water quality outcomes is to physically remove the pipe wall accumulations as would be practicable. In addition, the water must be controlled to prevent excessive growth of all microorganisms (not just pathogenic ones) and their biofilm formation. That is, the water must be made "biologically stable."

2.1 COMPARISON WITH THE REGULATORY PERSPECTIVE

In contrast to a comprehensive perspective, the drinking water regulations for distribution system water quality issues treat each issue separately. For example, in the Lead and Copper Rule, the release of lead and copper into the drinking water from piping materials is the focus of the regulation. The water entering the pipe is a solution of carbonate compounds. See Figure 2.1 and Table 2.1. There are no other chemicals assumed to be in the water and no microorganisms are assumed to be present. The pipe wall accumulations are hypothesized as only containing carbonates of lead or copper—again, no other chemicals or microorganisms. The quantity of lead or copper dissolved in the water is dependent on the solubility of the lead or copper carbonate compounds formed. The more soluble the compound, the more lead or copper dissolved in the water. From the regulatory perspective, lead and copper control is a matter of finessing the pH and/or alkalinity (somewhat equal to carbonate concentration) of the water to produce more insoluble compounds of lead and copper carbonates. Alternatively, orthophosphate can be added to form highly insoluble compounds of lead and copper phosphates. To understand orthophosphate chemistry, substitute the word "orthophosphate" in Figure 2.1 and Table 2.1 for the word "carbonate."

2.2   A CLOSER LOOK AT WHAT SHAPES DRINKING WATER QUALITY

The observations that comprise the comprehensive perspective of water quality have confirmation in the scientific literature. This section takes a closer look at several of these factors which shape drinking water quality. Readers who are anxious to begin investigating water quality can proceed directly to Chapter 3, titled "A Quick but Powerful Assessment of Drinking Water Quality." But investigations can be more effective and insightful with fundamental concepts in mind. Read further in Chapter 2 for fundamental concepts regarding water quality.

Factors that shape drinking water quality can be grouped into three major categories for organizational purposes:

* Uniform corrosion of metals
* Biostability of water
* Chemical scale formation and dissolution

2.2.1   Uniform Corrosion of Metals

Uniform corrosion of metals is an electrochemical interaction. It is the same process that occurs in a battery. That is, electrons flow through a metal pipe in the same way that they flow through a battery (AwwaRF and DVGW-TZW 1996; Peabody 2001).

Figure 2.3 shows the components of a battery that are essential for electrons to flow. There must be a negative terminal, called an "anode," to provide the electrons that make up electrical flow, and a positive terminal, called a "cathode," to receive the electrons. There must also be a medium, such as a wire, to allow the electrons to flow from the anode to the cathode.

In addition, there must be certain chemical characteristics of the battery that make electrons "want" to flow from the anode to the cathode. These characteristics are based on the structures of the atoms in the anode and the cathode.

An atom is made up of a nucleus with electrons orbiting it at various distances in a similar way that planets orbit the sun (Figure 2.4). Based on various orbit structures, some atoms have a tendency to freely give up electrons. Other atoms have

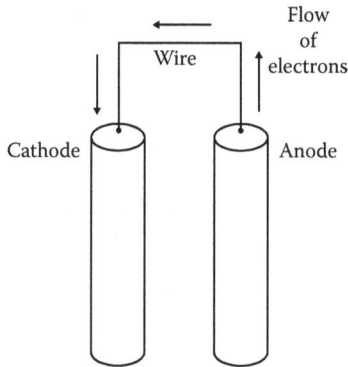

FIGURE 2.3 Main components of a battery.

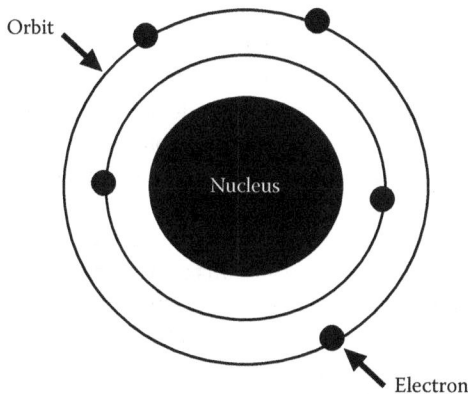

FIGURE 2.4 Atom with nucleus and electron orbits.

structures that allow them to accept electrons. If atoms with a tendency to give up electrons are connected by wire to atoms that are willing to accept electrons, the electrons will flow.

Consider, though, that atoms are neutral in electric charge. The negative charges of the electrons balance the positive charges from the nucleus. What happens when an atom loses one or more electrons? With some electrons missing, the atoms now have more positive charges than negative charges. They are called "positive ions" or "cations." The effect of this is that the neutral solid metal atom dissolves as a cation into the surrounding environment. Because solid metal is lost at the anode in this process, the metal is said to "corrode."

Nature does not like this excess charge and requires "electroneutrality," where positive charges must be paired with equal negative charges. The creation of a positive ion can only occur if there is a negative ion ("anion") of equal charge with which to pair. The metal ion finds a mate in the surrounding environment and a new compound is formed.

Slightly different chemistry occurs at the cathode. The cathode atoms receive the electrons from the anode but do not necessarily need to incorporate them into their atomic structure. The atoms are able to pass the electrons on to a willing electron acceptor in order to achieve electro-neutrality. There must be chemicals in the environment that surround the cathode, accepting the electrons, achieving electroneutrality, and forming new compounds.

Figure 2.5 shows an example of chemical interactions in a battery where a zinc anode and a carbon cathode are used. Sulfuric acid surrounds the electrodes. As the chemical reactions in the battery proceed, the zinc anode corrodes, forming dissolved zinc sulfate. The cathode passes electrons on to hydrogen cations, forming hydrogen gas.

The same interactions can occur between a metal pipe and adjacent water. When uniform corrosion occurs in pipes, the anode and the cathode are located on the surfaces of the metal pipe. More specifically, there are many microscopic anodic and cathodic sites on the pipe wall and these sites move around randomly (Figure 2.6). This dynamic change of locations is thought to occur because of microscopic imperfections and variations on the metal surface (AwwaRF and DVGW-TZW 1996, p. 14).

Wherever metal atoms lose electrons, solid metal is lost and the positive metal ions go into solution with the adjacent water. Because the anodic sites move around randomly, there is an equal chance for any site on the metal pipe to corrode. Solid metal is lost uniformly along the metal pipe surface. This is called "uniform corrosion" (Figure 2.7).

The "wire" that carries the electrons from the anodic sites to the cathodic sites is the metal pipe itself. It is hard to envision this as a simple battery, because the metal pipe serves as all three of the battery's solid components.

The water adjacent to the metal pipe provides all the other chemical components of the battery. Drinking water is not just pure H_2O composed only of hydrogen and oxygen. Water carries many dissolved chemical compounds and can provide electron acceptors to the cathodic sites at the pipe wall. Water adjacent to anodic sites on the pipe wall can provide negatively charged ions to pair with the newly dissolved positive metal ions.

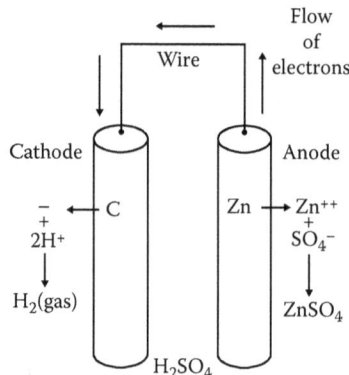

FIGURE 2.5   Example of chemical interactions in a battery.

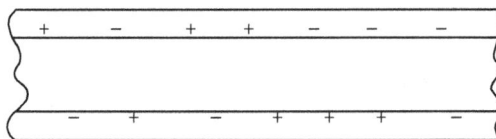

FIGURE 2.6 Metal pipe/water battery: anodic and cathodic sites.

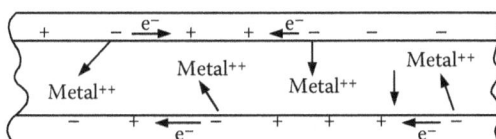

FIGURE 2.7 Metal pipe/water battery: uniform corrosion.

The formation of new compounds is a significant step in the corrosion process of the metal pipe/water battery. The newly formed compounds ("corrosion by-products") have their own specific chemical and physical properties. Some compounds will be highly soluble in water. That is, even when a high concentration of the compound is present, it remains dissolved in the water. Other compounds will have a low solubility in water. That is, after a small concentration of the compound is formed, any excess quantity drops out of the water as a solid. Depending on the size and other properties of the solid's molecules, the compound will form a fine film to a coarse scale on the pipe wall. The buildup on the pipe wall will determine if uniform corrosion will proceed, slow down, or cease.

More specifically, to control corrosion inside water pipes, the flow of electrons needs to be stopped. In a common battery, the flow of electrons and the subsequent corrosion of the anode can be prevented by disconnecting the wire between the anode and the cathode; an on/off switch breaks the wire. Unfortunately, there is no switch within a metal pipe; a different strategy must be used. That strategy is to prevent the chemical interactions from occurring at the interface between the water and the pipe wall. For example, if there is not an electron acceptor at the cathode or an ion mate for the metal at the anode, the electrons cannot flow. However, it is not always practical to control the composition of the water.

Another way to prevent the chemical interactions is to create a barrier on the pipe wall. The barrier must uniformly cover the metal surface and must be nonporous to prevent dissolved ions and molecules from reaching the metal surface (Figure 2.8).

Many corrosion by-products build up on the pipe wall, but not all corrosion by-products form solids that meet the criteria for an effective barrier. Common corrosion

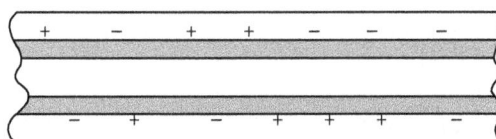

FIGURE 2.8 Metal pipe/water battery: nonporous barrier preventing electron flow.

by-products that form naturally in water are metal carbonates, hydroxides, and oxides. Only when the corrosion by-product is highly insoluble in water and coats the pipe wall with a uniform, nonporous film does uniform corrosion slow or stop. When this happens, the water is referred to as "noncorrosive" or "nonaggressive."

If the water flowing through the pipes does not have naturally occurring chemicals that can form protective films, chemical products can be added to water to provide these characteristics. An example of such a corrosion inhibitor is orthophosphate. The phosphate ions pair with metal ions, such as lead or copper, to form highly insoluble metal phosphate compounds. These compounds precipitate onto the pipe wall as an insoluble uniform, nonporous film and corrosion is slowed or stopped.

This simplified discussion of batteries and chemical reactions can be summarized in two general concepts that define uniform corrosion in water systems. One is that the whole phenomenon happens because of a battery-like flow of electrons. The second is that, in pipes, the nature of the corrosion by-product compounds controls the fate of the corrosion process itself.

The Lead and Copper Rule is an example of the use of uniform corrosion concepts as was discussed earlier in reference to Figure 2.1 and Table 2.1. It is focused on the solubility of carbonate compounds based on the fact that carbonates, typically found in water, can form compounds with lead and copper ions after they are released from piping materials.

Alternatively, the Lead and Copper Rule considers that orthophosphate can be added to form highly insoluble compounds of lead or copper phosphates. Then, the pipe wall accumulations are composed of orthophosphate compounds, which are more insoluble than lead or copper carbonate compounds. As stated previously, Figure 2.3 can be redrawn with phosphates of lead or copper instead of carbonates.

However, there are other metal compounds that can form as corrosion by-products in the uniform corrosion process outside of the Lead and Copper Rule's consideration. For example, oxides of lead have been found to typically be present on pipe walls. In addition, the presence of a more insoluble form of lead oxide has been acknowledged as contributing to lead control in water systems (Lytle and Schock 2005). Also, chlorides and sulfates have been acknowledged as creating more soluble lead and copper compounds, helping to perpetuate the corrosion of the pipe walls (Larson and Skold 1958; Nguyen et al. 2010, 2011).

2.2.2 Microorganisms and the Biostability of Water

A second category of water quality factors concerns the biostability of water, where the growth of microorganisms is to be kept in check. The role of microorganisms in shaping the water quality continues to be a misunderstood topic. Microorganisms, ever-present in drinking water, are typically approached as consumer health and safety risks. Not all microorganisms pose such risks, but have the potential to degrade water quality and even damage pipes.

One type of water quality degradation that can be caused by microorganisms is metal corrosion resulting in increased metal concentrations in the water and/or pinhole leaks in pipe walls. However, the viewpoint of many drinking water researchers and practitioners appears to assume that metal corrosion occurs first, caused solely by chemical factors, and this creates debris where the microorganisms take up residence. This is exemplified in a study of biostability where corrosion rate was studied as a factor contributing to the formation of biofilms instead of formation of biofilms contributing to the corrosion rate; causality was not proven, it was assumed (LeChevallier et al. 2015). An article on iron pipe corrosion stated, "a diverse microbiological community (sic) can be found in the scale" (Burlingame et al. 2006). The article described the complete corrosion process as chemical only, with the corrosion debris providing a structure to which microorganisms attach. Both of these articles imply an assumption that microorganisms have no role in the corrosion process.

The reality is that several aspects of microbiological growth can cause corrosion of metal surfaces and solubilization of metal compounds. It is known that microorganisms can secrete acidic enzymes to attach to metal surfaces (Bremer et al. 2001) and that such localized acidity can corrode the surfaces.

It is also known that microorganisms can produce acidic waste products, such as hydrogen sulfide, from sulfate-reducing bacteria, which forms a weak acid in water (Rittman and McCarty 2001; Madigan and Martinko 2006), another pathway to increased metal corrosion.

In yet another pathway, microbiological waste products, such as nitrates and acetates, can form highly soluble compounds of lead and copper and can possibly resolubilize existing lead and copper films on metal surfaces. The waste products can also work directly in the uniform corrosion electrochemistry to produce highly soluble corrosion by-products (see Section 2.2.1).

It is also known that there are iron-oxidizing bacteria that use electrons from iron and other metals as their food source (Rittman and McCarty 2001; Madigan and Martinko 2006), another pathway by which metal can be oxidized by microorganisms in a water system.

From the above list of possible microbiological pathways that can affect metal corrosion, it can also be theorized that microbiological life cycles can produce both particulate metals and dissolved metals.

Particulate metals can occur when electrons are utilized by microorganisms directly from metals, as described for iron-oxidizing bacteria, and oxidized metals in the form of solids are produced. Particulate metals can also result if biofilms on pipe walls experience a die-off in a changing water environment. With biofilms intertwined with chemical scales as have been observed on water system surfaces, chemical scales can be broken up or metals adsorbed to biofilm material can be released to the water.

Dissolved metals can occur when acidic waste products are released into the water, lowering the general water pH. Dissolved metals can also occur with microbiological waste products, such as nitrates and acetates, that form highly soluble compounds with metals, such as lead and copper.

Microorganisms can cause other water quality changes besides metal corrosion:

- Disinfection residual can be depleted by an excessive population of microorganisms, leaving the water system vulnerable to microorganisms, such as the bacteria *E. coli*, which can enter the water system through main breaks or cross-connections and make people ill.

- Microorganisms can form biofilms on pipe walls that can harbor hosts of *Legionella*, a bacterium that in some cases can cause a deadly respiratory illness.
- Microorganisms can produce acetate as a waste product that forms carcinogenic compounds with chlorine disinfection.

The key to lowering the potential for microbiologically influenced water quality changes is to keep the microbiological populations in balance. Factors that encourage the growth of microorganisms must be balanced against factors that discourage their growth. Successful balancing of factors is called "biostability" (van der Kooij 1992; Volk and LeChevallier 2000; Zhang et al. 2002; LeChevallier et al. 2015).

The concept of biostability also leads to the realization that microbiologically influenced corrosion and other negative water quality effects from microorganisms are systemic in a distribution system. A colony of microorganisms in a biofilm may be localized, but its microbiological nutrients, microbiological waste products, and microorganisms themselves migrate and can be measured throughout a distribution system. In this way, the potential for microbiological life cycles with resultant metal corrosion and other negative effects pervades a distribution system.

2.2.3 CHEMICAL SCALE FORMATION AND DISSOLUTION

The Lead and Copper Rule and its uniform corrosion model only include dissolved lead and copper release and not particulate lead and copper. But chemical scales have been acknowledged as a complexity that shapes lead and copper release in the distribution system (Schock et al. 2014). Chemical scales, such as iron, manganese, and aluminum, can adsorb other metals such as lead, copper, arsenic, and radium, accumulate them, and release them as significant particulate concentrations when the coarse scales crumble. Or, if the water environment, such as the pH of the water, changes, the chemical scales and their adsorbed metals can redissolve in the water.

Existing chemical scales also can physically block and prevent the lead or copper carbonate, oxide, or phosphate solids from forming a uniform, nonporous barrier on pipe walls (DeSantis and Schock 2014).

2.3 SUMMARY

In summary, a comprehensive perspective of drinking water quality is required in order to work with the complexity of water distribution systems. The comprehensive perspective attempts to account for all factors that can shape water quality in a distribution system. In this attempt, factors are sorted into three main categories:

- Uniform corrosion of metals
- Biostability of water
- Chemical scale formation and dissolution

These three main categories will be mentioned frequently in this book.

This approach contrasts with the simpler perspective used by drinking water regulations. The Lead and Copper Rule is an example of an approach where each distribution system contaminant is thought of as isolated from other distribution system contaminants; the reasons for the production of the contaminant in the distribution system are simplified and idealized. It is easier to regulate simpler concepts and measurements.

However, working with the comprehensive perspective of water quality does not have to be overwhelming and expensive. Chapter 3 describes four relatively easy steps that can routinely indicate the status of the distribution system water quality.

3 A Quick but Powerful Assessment of Drinking Water Quality

In four relatively easy steps, water utility personnel can know if the distribution system has the potential for poor-quality water and even indicate if it is beginning to trend in that direction.

3.1 WATER SYSTEM CONFIGURATION AND OPERATIONS PARAMETERS

The first step is to summarize information that describes the water system's configuration and its operation. It is good to have this information available and updated annually because it can be used for many aspects of engineering planning. In terms of the comprehensive perspective of water quality, the configuration information can hold clues to system operational changes that may affect the water quality.

One example of important system information involves the quantity of water used from one year to the next. If there is a large drop in water usage one year, then the water age (residence time) of the water in the system will have increased. This can create an environment for microorganisms to grow excessively, with negative effects on water quality, as was discussed under Section 2.2.2.

In another example, one water system found a dramatic increase in unaccountable water loss one year, with the increase continuing in the next year's data. This motivated a leak-detection program that uncovered an epidemic of pinhole leaks in copper water service lines (Cantor, Bushman, Glodoski 2003; Cantor et al. 2006).

In addition, changes in water sources, entry points to the distribution system, water treatment processes, water treatment chemicals, and water storage can all cause changes in water quality.

It is also important to know the water system's inventory of materials of construction with focus on water transmission lines, water mains, and water service lines. Materials of construction can contribute metals and other chemicals to the adjacent drinking water.

All of this information is compiled in most water systems for various operational and financial reasons. It is equally important to keep this information organized for the sake of understanding water quality changes and to record how the water system configuration and operations parameters change from year to year.

Important information to summarize is as follows:

- Population served
- Number of customers

- Water audit
  - Total pumpage
  - Wholesale customer sales
  - Retail customer sales
  - Water used for flushing, fire protection, and pipe-freezing protection
  - Piping breakage loss
  - Leakage and other unaccountable loss (total pumpage minus sales and estimated known losses listed above)
  - Percent of unaccountable water loss (leakage and other unaccountable loss/total pumpage)
- Water sources
  - Main
  - Emergency
- Entry points to the distribution system
  - Main
  - Emergency
- Water treatment processes and chemicals
  - Treatment plant name
  - Intake chemicals, if relevant for surface-water plants
  - Primary disinfection methods, if relevant for surface-water plants
  - Water treatment processes
  - Water treatment chemicals used in the processes
  - Secondary disinfection methods
  - Corrosion control chemicals
  - Other chemicals
- Water stored in the system
  - Volume of water stored in standpipes, reservoirs, and elevated tanks
  - Number of storage facilities
- Water transmission line inventory
  - Type of material
  - Diameter of pipes by material
  - Length of pipes by material in feet
  - Annual repairs
- Water main inventory
  - Type of material
  - Diameter of pipes by material
  - Length of pipes by material in feet
  - Annual repairs
- Water service line inventory
  - Type of material
  - Diameter of pipes by material
  - Number of service lines by material
  - Annual repairs

Other information relevant to specific systems can be added to this list. The information can be presented in several ways in order to think about the information in different contexts.

One method is to tabulate the information by year in order to visually scan for changes in the information from year to year. In Table 3.1, water system data are tabulated over a four-year period.

TABLE 3.1

Example of Water System Configuration and Operations Parameters

	Population Served		
Year	Inside Municipality	Outside Municipality	Customers
2012	104,000	0	36,092
2013	104,000	0	36,164
2014	104,000	0	36,164
2015	104,000	0	35,742

	Water Sources		
Time Period	Groundwater	Surface Water	No. of Entry Points
2012–2015	9 wells for emergency use	Lake water	9 wells for emergency/2 lake

Plant Name	Intake Chemicals	Primary Disinfection	Treatment	Treatment Chemicals	Secondary Disinfection	Corrosion Control Chemicals	Other Chemicals
Central Treatment Facility	Chlorine gas	Ozone	Flocculation, sedimentation, sand filtration	Polyaluminum hydroxychloride as coagulant, carbon dioxide, sodium bisulfite	Sodium hypochlorite	None	Fluoride

	Annual Pumpage in 1,000 gallons				Water Loss in 1,000 gallons	
Year	Ground	Surface	Purchased	Total	Total Non-Revenue	Unaccountable
2012	561	6,689,201	337	6,690,099	297,100	198,715
2013	759	6,360,570	312	6,361,641	365,830	173,871
2014	3,548	6,500,494	297	6,504,339	549,861	222,659
2015	1,054	6,496,540	451	6,498,045	525,370	414,492

(Continued)

TABLE 3.1 (CONTINUED)
Example of Water System Configuration and Operations Parameters

	Water Mains (Miles by Type)				Service Lines (Number by Type)		
Year	Metal	Plastic	Lead	Other	Metal	Plastic	Lead[a]
2012	423.6	47.8	0.02	41.8	30990	2133	2472
2013	418.9	50.3	0.02	41.8	30857	2324	2356
2014	417.2	51.2	0.02	41.8	30858	2429	2337
2015	418.9	48.5	0.02	41.8	30742	2609	2251

Year	No. of Water Main Breaks	No. of Service Line Breaks
2012	127	22
2013	172	8
2014	290	32
2015	147	33

Time Period	Storage Facilities					
	Reservoir (gallons)		Elevated Tank (gallons)		Standpipe (gallons)	
2012–2015	12,000,000	7 tanks	3,300,000	4 tanks	2,000,000	1 tank

[a] Most lead services in this system are connected to privately owned galvanized iron service lines.

Tabulated data are easy to graph to better visualize changes over time for each type of information. For example, Figure 3.1 shows that water pumpage dropped in 2015. Unaccountable water loss increased in 2015. This would be useful information for financial considerations, but it could also be significant in uncovering reasons for changes in water quality.

A simplified system map is another good way to show features of a water system—geographical location of water sources, water treatment, and storage tanks.

3.2 WATER SYSTEM TIMELINES

Annually summarizing the water system configuration and operations information becomes somewhat of a diary that may hold clues to water quality changes. There still may not be enough detail to pinpoint why water quality changes occurred. This is why Step 2 of the water quality assessment is to create timelines of events in a water system.

The timeline should start in the past. Those entries will highlight major changes that occurred in the history of the water system, such as noting major construction or operational changes. Table 3.2 is an example of a water system that began as a groundwater system. Then, the system was eventually switched to a surface water source. In more recent years, upgrades were made to the treatment plant.

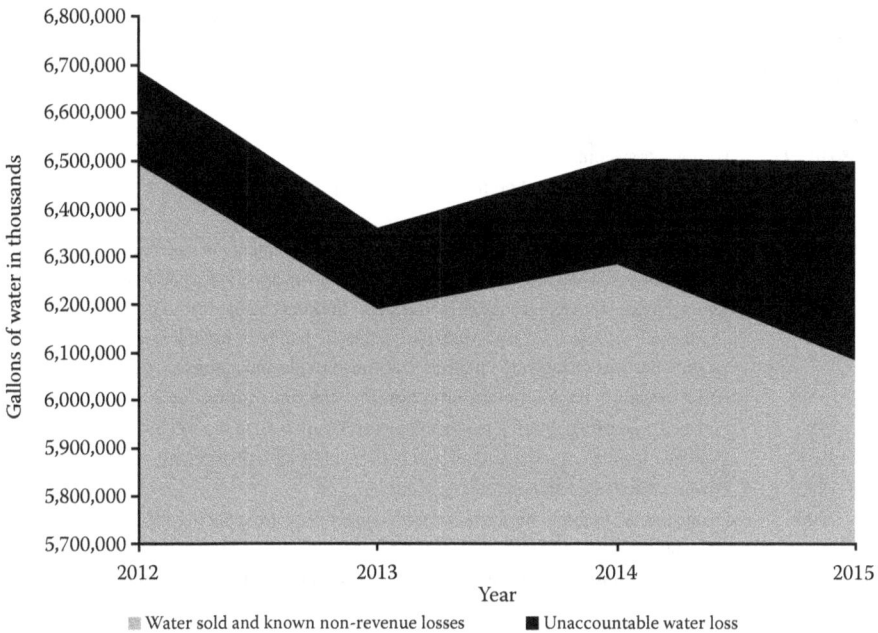

FIGURE 3.1   Example of graphed water audit data.

As a timeline becomes more current, other details—chemical product changes, chemical dosage changes, major or systemic customer complaint events, regulatory compliance issues—can be added. See Table 3.2.

A detailed daily log at a water utility could be kept in a computer spreadsheet or database. Each department in the utility, such as General Management, Accounting, Customer Service, Water Quality, Water Treatment, and Distribution, could keep its own log with all logs viewable by other departments. Table 3.3 shows examples of event log entries.

Log entries can be less detailed than described in Table 3.3, but some degree of recording of operations events should be performed.

3.3   LEAD AND COPPER RULE DATA

Lead and Copper Rule compliance data are quite helpful in pinpointing time periods when general water quality might have been challenged. Studying the data aids in relating lead and copper release and other distribution system issues to operational changes.

Table 3.4 is an example of lead data from one Lead and Copper Rule compliance sampling event from a water utility. The Rule can require two monitoring periods a year—(1) January through June and (2) July through December—so the sampling event can be identified by year and semester, such as 2017-2, which is sampling performed after June in 2017.

Also shown in Table 3.4 are statistics for the dataset. A statistic used for Lead and Copper Rule compliance is the 90th percentile concentration. A 90th percentile

TABLE 3.2
Example Timeline for a Water System

Date Range	Event
1886 to 1957	Well water system.
1955	Raw water transmission line from lake to filter plant built.
1956	Intake 1 built.
1957	Raw water pumping station and conventional filtration plant are operational. Surface water now used as the water source, with groundwater as backup source.
1972	Various treatment changes made by this year: fluoride changed from powder to liquid hydrofluorosilicic acid; potassium permanganate had been used at raw water pumping station but was changed to chlorine; chlorine dioxide and ammonia were used for brief periods at the filter plant but is now free chlorine disinfection.
1990	Finished water free chlorine residual increased from 0.40 to 0.55–0.60 ppm.
1991	Switched from alum to polyaluminum hydroxychloride as coagulant.
1992	Powdered activated carbon feed stopped.
2000	Ozone online; begin to have trouble with copper pressure reducing valve components; now all are replaced with stainless steel.
2005	Second transmission line from lake to plant was built and in operation by June; treatment plant capacity was increased; a sodium hypochlorite system was installed to replace the use of gaseous chlorine.
2006	Switched from chlorine gas to sodium hypochlorite.
2007	Finished water free chlorine residual increased from 0.6 to 0.70–0.75 ppm.
2008	March: A carbon dioxide injection system was installed in the sodium hypochlorite carrier water to prevent scaling in the solution delivery line.
2010	Variable frequency drives installed at lake water pumping station.
2011	A section of one transmission line was replaced.
2011	Filtration reservoir leak investigation.
2011	Distribution system modeling.
2011	Fluoride residual lowered to 0.70–0.80 ppm.
2011	Lead and Copper Rule compliance: exceeded lead action level.
2013	Break in 36-inch main.
2014	January, February: severe winter weather causing increased main leaks.
2014	January, February: severe winter weather causing increased frozen services.
2014	Summer: first year of unidirectional flushing.
2015	Summer: second year of unidirectional flushing.

lead concentration means that 10% of the lead concentrations measured during a sampling event were higher than that concentration and 90% were equal to or lower. When a water system has 90th percentile lead and copper concentrations below the regulatory Action Level (15 µg/L for lead and 1300 µg/L for copper), it is deemed in compliance with the Rule. Then, monitoring is required once every three years in the warmer semester, Monitoring Period 2. If out of compliance, a water system must demonstrate a 90th percentile lead or copper concentration less than the published Action Level for two semesters in a row and again in a semester a year later.

TABLE 3.3

Example Water Department Event Logs

Utility:	Acme City Water Utility
Department:	Distribution

Date	Log Entry
8/22/17	Well 4 chlorine dosing pump motor overheated. Has probably been out of service 24 hours.
8/23/17	Still waiting for delivery of pump part.
8/24/17	Received pump part.
8/25/17	Installed pump part. Dosing pump working again.

Utility:	Acme City Water Utility
Department:	Customer Service

Date	Log Entry
8/22/17	3 complaints of discolored water in Well 4 service area.
8/23/17	12 more calls.
8/24/17	15 more calls.
8/25/17	17 more calls.

Utility:	Acme City Water Utility
Department:	General Manager

Date	Log Entry
8/22/17	
8/23/17	
8/24/17	Call from mayor's office about discolored water.
8/25/17	Call from newspaper reporter.

The 90th percentile concentrations plotted over time can be useful in visualizing water quality trends along with other statistics shown in Table 3.4—the minimum, median (50th percentile), average, and maximum values.

Statistics from another water utility of Lead and Copper Rule compliance lead data starting from 1992 are plotted on a graph in Figure 3.2. Only the 90th percentile data were available before 2000. After 2000, the other statistics for the sampling datasets are shown. The 90th percentile line shows an increasing trend in lead concentration starting in 2000. In 2011, there was a large jump in lead concentrations that pushed the water system out of compliance with the Lead and Copper Rule. (The maximum was so high that year that it is not shown on the graph.)

In looking back at Table 3.2, which is a timeline for this water system, it is seen that an ozone process was installed in 2000. A second large transmission line was installed in 2005 and the first transmission line was repaired in 2011. Further study of

TABLE 3.4

Example of Lead Data from One Lead and Copper Rule Sampling Event

2016 Semester 2 Lead Data in µg/L Sorted	Statistics
0.00	Minimum = 0.00
0.10	
0.11	
0.11	
0.11	
0.12	
0.13	
0.14	
0.15	
0.17	Median (50th percentile) = 0.17
0.17	
0.18	
0.20	Average = 0.25
0.32	
0.34	
0.36	
0.43	
0.56	90th percentile = 0.56
0.59	
0.68	Maximum = 0.68

these features and existing data showed how the activities during those years helped to push the water system out of compliance with lead. Disinfection by-products also increased in the water system, and there was an increase in corrosion of parts in meters and small valves. The case will not be discussed in further detail here, but it is important to note how the study of the Lead and Copper Rule data with the timeline directed investigators to study the particular features that were added just before the water quality changed.

In another water system, copper went out of compliance in the 2008 Lead and Copper Rule compliance sampling. See Figure 3.3.

The Lead and Copper Rule data, combined with the water system's timeline, quickly identified a new well placed online one year before the Lead and Copper Rule compliance issue was found as the potential source of several water quality issues. In an investigation that followed, the new well was found to have several problems—microbiologically influenced corrosion occurring in the well, elevated iron and manganese in the well water, and elevated arsenic from the aquifer. These translated into several issues in the distribution system that affected consumers—elevated copper, elevated arsenic, and discolored water.

These are examples of how informative Lead and Copper Rule compliance data graphs used in conjunction with the water system configuration and operations

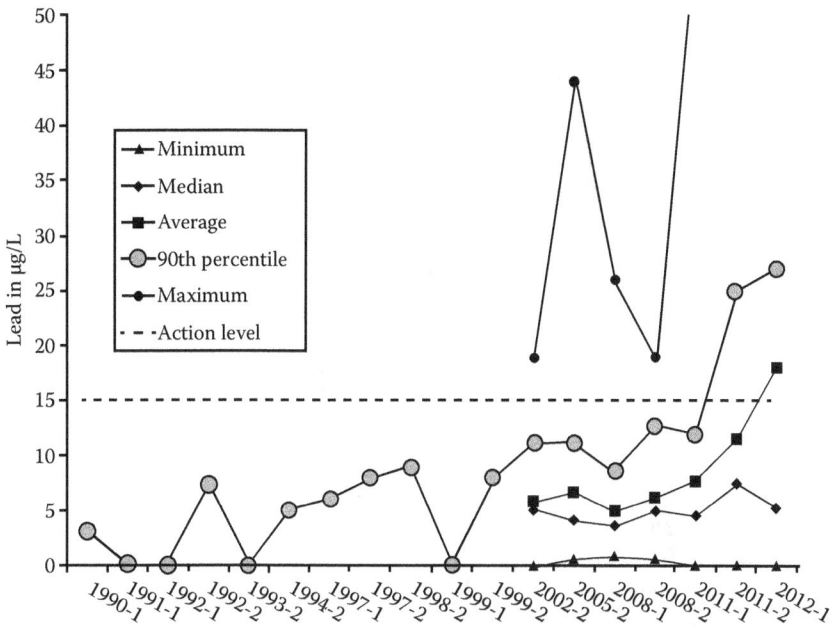

FIGURE 3.2   Example plot of historical utility Lead and Copper Rule data: Utility 1 lead data.

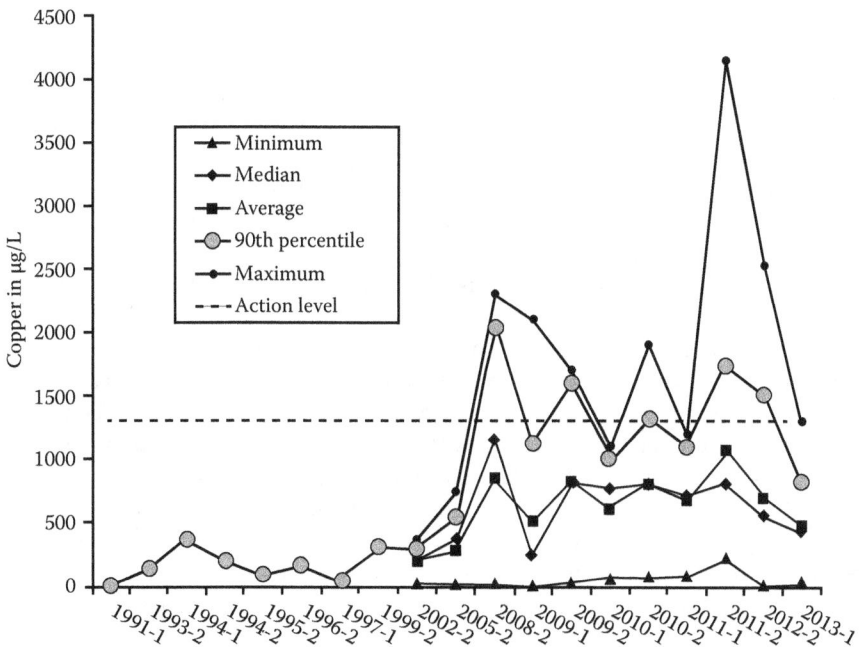

FIGURE 3.3   Example plot of historical utility Lead and Copper Rule data: Utility 2 copper data.

parameters and timelines can be in locating specific water system operations that may have affected not only lead and copper release but also other water quality characteristics. Studying Lead and Copper Rule data is Step 3 of the quick but powerful assessment of drinking water quality.

3.4   TOTAL COLIFORM RULE SITE PARAMETERS

Step 4 of the water quality assessment provides data on a more routine basis than the Lead and Copper Rule data. The data come from established distribution system sampling sites.

The Total Coliform Rule, one of the federal drinking water regulations, conveniently requires water utility personnel to select flowing water sampling sites around the distribution system to represent geographical changes in water quality. Sites are visited weekly. Other already-established distribution system regulatory sites or utility operational sampling sites can be used for this data-gathering effort as well.

At these distribution system sites, two easily accessible and economical field test parameters that represent the comprehensive water quality perspective can be measured routinely. That is, from Chapter 2, the comprehensive water quality perspective implies that chemical scale and biofilm accumulations on pipe walls should be removed in order to lower the potential for negative water quality changes in the distribution system. In addition, the potential for excessive growth of microorganisms in the water system must be lowered. So, water quality parameters that represent the absence of particulate matter and microbiological activity in the water are desired.

Disinfection concentration is one of the parameters. When disinfection concentration is low at a location in a water distribution system, it means that the chemical has been used up. This is especially evident when the concentration is compared to the original dose at the entry points to the distribution system. The disinfection can be depleted by interactions with existing pipe material and pipe wall chemical scale compounds or by interactions with existing microorganisms and biofilms in the system as well as bulk water chlorine demand compounds. Disinfection depletion is also a function of water age (residence time) in the water system with longer exposure to disinfection-depleting factors. If pipe walls are clear of chemical and microbiological accumulations, disinfection concentration can be found to be at similar levels around the distribution system as was freshly dosed at the entry points to the distribution system. In this way, disinfection concentration is an indicator of the chlorine demand of the water and pipe wall accumulations and an indicator of the cleanliness and biostability of the water system. Although disinfection needs vary, anecdotally, if concentrations are below 0.2 to 0.3 mg/L free chlorine for chlorine disinfection or below 1 mg/L total chlorine for chloraminated systems, there are most likely not adequate levels of disinfection to defend against excessive growth of microorganisms. Disinfection concentration is one of the parameters required to be measured at weekly visits to Total Coliform Rule sites. Most water utilities have years of historical disinfection data from those sites and continue to gather this information weekly.

Another easily obtainable water quality parameter is turbidity. Turbidity is a measure of particulates entrained in the water. Particulates can represent inert material such as sand, chemical scale particles such as particulate iron or manganese, microorganisms, or biofilm materials. In this way, turbidity is another indicator of water system cleanliness and biostability. Turbidity should not exceed 1 NTU in the distribution system based on a study of the potential to form biofilms in water systems (LeChevallier et al. 2015). Anecdotally, negative water quality effects can potentially begin when the turbidity exceeds 0.5 NTU. The measurement of turbidity can be added to routine distribution system sampling site visits.

Tracking disinfection concentrations and turbidities routinely at set locations around the distribution system is equivalent to tracking the status of cleanliness and biostability over time and location. It is equivalent to routinely tracking the potential to develop negative water quality characteristics in the distribution system as a function of location and time. With this routine information, problem locations or operational periods can be pinpointed and remedied on a routine basis.

As an example, data are presented from one water system that uses chloramine disinfection. In this case, total chlorine concentration represents the active disinfection in each water sample. (Systems that use free chlorine disinfection should measure and track free chlorine concentration as the active disinfectant concentration measured at each site.) Table 3.5 displays the example disinfection data written as total chlorine concentration in mg/L.

A time period of study must be selected from the data before graphing. For example, quarterly data coincide with seasons. Changes in water quality by season can be studied when using these data. However, the number of data points in a time period must also be considered. The fewer the data points available in a time period, the less accurate the representative statistics are.

TABLE 3.5
Example of Total Coliform Rule Site Disinfection Data

Sample Site	1	2A	3	4	...	15	16
11/15/2015	2.20	2.80	1.70	2.10	...	2.10	2.30
11/9/2015	2.40	2.60	2.00	2.30	...	2.10	2.50
11/2/2015	1.73	2.40	1.40	2.00	...	1.62	1.67
10/20/2015	2.01	2.30	1.74	2.10	...	1.81	2.11
10/11/2015	1.41	1.80	1.32	1.32	...	1.50	1.41
10/4/2015	1.60	2.10	1.40	1.89	...	1.60	1.90
9/20/2015	1.93	2.50	1.89	2.30	...		2.40
9/14/2015	1.43	2.30	1.38	1.86	...	1.52	1.82
9/7/2015	1.30	1.90	1.20	1.70	...	1.70	1.60
...	...	...	...	...	...	...	...
12/8/2009	1.35		1.39	1.68	...	1.57	1.66
11/17/2009	1.31		1.39	1.68	...	1.50	1.53
11/12/2009	1.48		1.48	1.84	...	1.74	1.79

Note: All data are total chlorine concentrations in mg/L.

In this example, the data from one water system are studied based on the first (Semester 1) and second (Semester 2) six months of the year. Data are shown in Table 3.6. For each semester of sampling, the average and the minimum and maximum data values are calculated. (Minimum and maximum values for each dataset will be used for now to represent the extent that each dataset varies. In Section 5.6, a specific statistical calculation of variation will be described and that range can be used instead of minimum and maximum in order to show where 99% of the values are expected to fall.)

For disinfection data in Table 3.6, all site statistics can be compiled into a graph. The statistics used are the average disinfection concentration at each site and the variability of results at each site. The sites are sorted by increasing averages so that the sites with the highest and the lowest averages are evident. The same data analysis and graphing can be performed for turbidity data at each site. See Figure 3.4 for a comparison of site disinfection data statistics for 2014 and 2015 in six-month time periods. (Site names are typically shown on the *x*-axis but are not shown in Figure 3.4 for simplification.)

Figure 3.4 shows the average concentration and variation of total chlorine at each Total Coliform Rule sampling site. Each square represents the average concentration at each sampling site. The "whiskers" extending from each square represent the

TABLE 3.6
Example of Total Coliform Rule Site Disinfection Data
for 2015 Semester 2

Sample Site	1	2A	3	...	14	15	16
11/15/2015	2.2	2.8	1.7	...	1.9	2.1	2.3
11/9/2015	2.4	2.6	2.0	...	1.9	2.1	2.5
11/2/2015	1.7	2.4	1.4	...	1.4	1.6	1.7
10/20/2015	2.0	2.3	1.7	...	1.4	1.8	2.1
10/11/2015	1.4	1.8	1.3	...	1.4	1.5	1.4
10/4/2015	1.6	2.1	1.4	...	1.3	1.6	1.9
9/20/2015	1.9	2.5	1.9	...	1.3		2.4
9/14/2015	1.4	2.3	1.4	...	1.1	1.5	1.8
9/7/2015	1.3	1.9	1.2	...	1.5	1.7	1.6
8/17/2015	1.6	1.9	1.1	...	0.8	1.7	1.9
8/9/2015	1.9	2.5	1.8	...	1.9	2.2	2.1
8/3/2015	1.4	1.8		...	0.8	1.7	1.8
7/18/2015	1.9	2.3	1.0	...	1.2	1.8	1.9
7/12/2015	1.8	2.3	1.6	...	1.1	2.0	2.1
7/5/2015	1.9	2.4	1.5	...		1.9	1.9
Minimum	1.3	1.8	1.0	...	0.8	1.5	1.4
Average	1.8	2.3	1.5	...	1.4	1.8	2.0
Maximum	2.4	2.8	2.0	...	1.9	2.2	2.5
Count	15	15	14	...	14	14	15

Note: All data are total chlorine concentrations in mg/L.

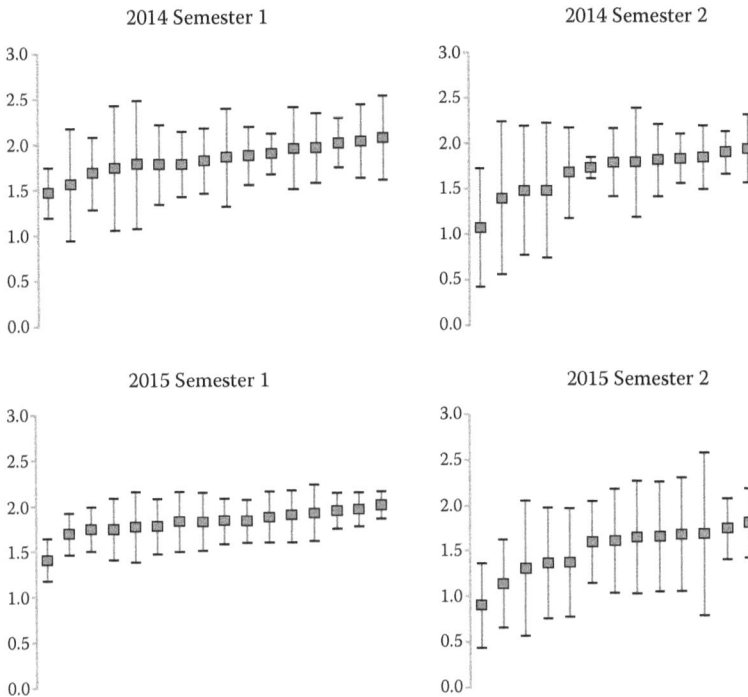

FIGURE 3.4   Comparison of total chlorine in mg/L at Total Coliform Rule sites.

variation of the data at each site. Range endpoints can be minimum and maximum values at each site, or they can be the result of a statistical calculation where 99% of the concentrations are expected to fall, as described in Section 5.6.

From this graph, sites with the lowest disinfection concentration can be easily identified and investigated. Sites with the highest variability can be identified and investigated to determine how to achieve a steadier disinfection concentration. Sites with disinfection concentrations below 1 mg/L as total chlorine for this chloraminated water system can be investigated to determine why the disinfection falls short of a minimum goal. In addition, it can be seen that there are lower average disinfection concentrations and more variability in concentrations in the second semester of each year, indicating a water quality change with seasonal temperatures. In this chloraminated system, a nitrification cycle with release of metals into the water in warmer temperatures was identified with further investigation. In this way, the disinfection data pinpointed specific sites and time periods where the potential for various water quality issues to occur increases.

The same analysis can be performed on turbidity data from the same sites, as shown in Figure 3.5. As with disinfection concentration, the worst-case sites can be identified and studied. For turbidity, it is desired to be under 1 NTU, preferably to stay under 0.5 NTU. Sites that exceed those goals, have the highest turbidity averages, or have the highest variation in turbidity are candidates for study and operational changes.

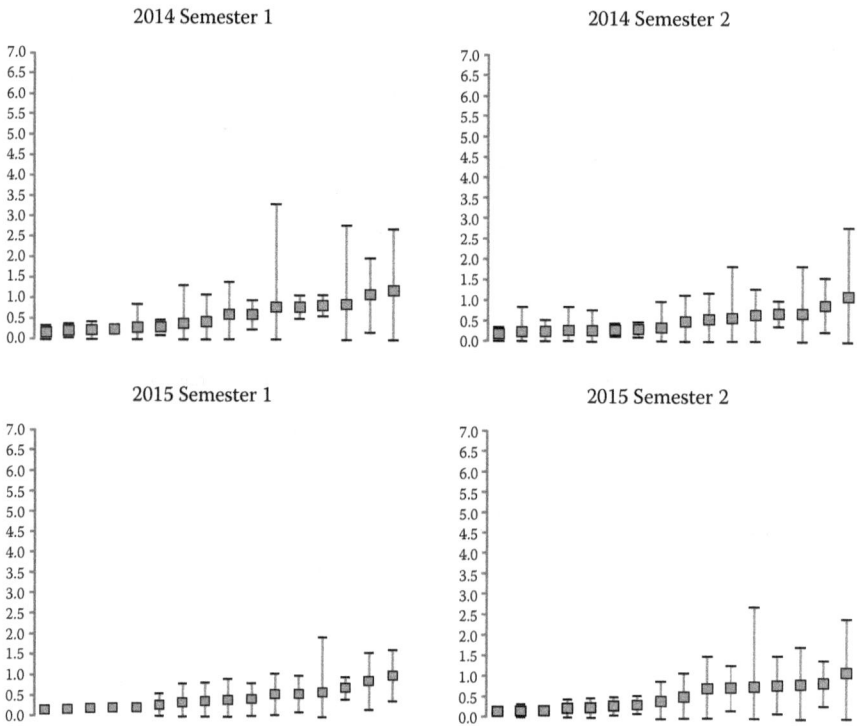

FIGURE 3.5   Comparison of turbidity in NTUs at Total Coliform Rule sites.

Additionally, all of the sites can be compared on a time-series graph together. See Figures 3.6 and 3.7. Each time-series line belongs to a sampling site but the sites are not identified on this graph. Instead, the graph is used to view the overall minimum and maximum results of all sites over the time period and to determine if all sites are exhibiting similar patterns.

In this example, the graphs of disinfection (Figure 3.6) show that some locations experience a drop in disinfection level during the peak nitrification period in September, October, and November. Other sites appear to be immune to this. With this knowledge, the sites with and without nitrification patterns, as identified from the disinfection graphs, should be compared to see if there is something operational that could be done to minimize or stop nitrification during the warmer-weather months at all sites. Finding the reason that some sites appear to be immune to nitrification could potentially lower lead and copper release and other distribution system water quality issues in the water system.

The turbidity graphs (Figure 3.7) offer the same opportunity to determine why there is great variation in turbidity at some sites in the water system and not at others. In this particular water system, it had been found by other means that 50% of the lead reaching consumers was in particulate form. Determining cleaning and operations routines that can lower the turbidity could protect consumers from sporadic contact with particulate lead in the drinking water.

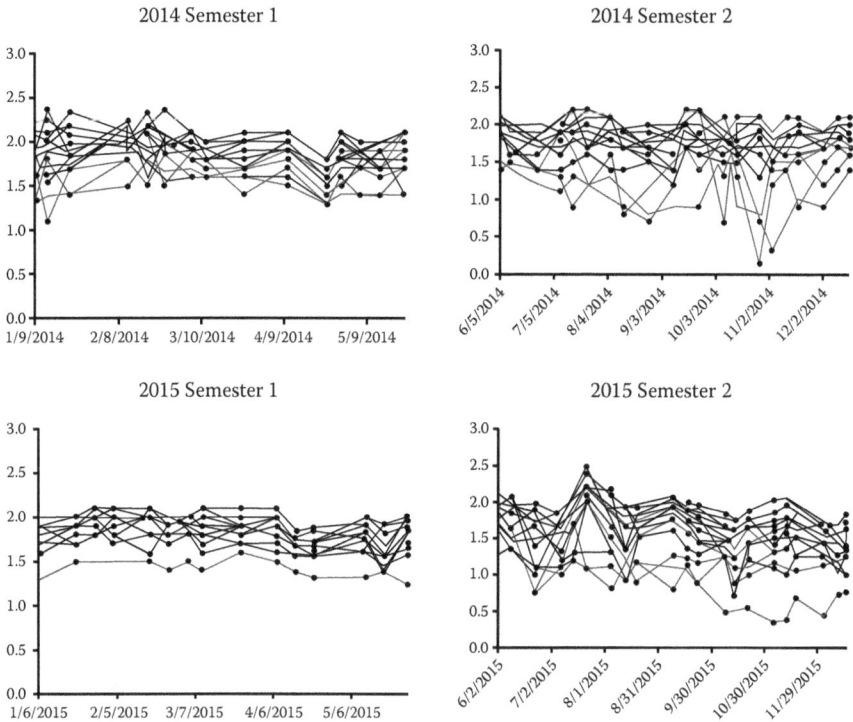

FIGURE 3.6   Comparison of total chlorine in mg/L at Total Coliform Rule sites over time.

In summary, disinfection concentration and turbidity data taken routinely at Total Coliform Rule sites and any other regularly visited distribution system sites can be used in conjunction with the water system configuration and operations parameters and the timelines to pinpoint operational events that may have resulted in reduced water quality. Locating sites that are immune to degraded water quality processes aids in making operational improvements at sites that show signs of water quality degradation.

3.5   SUMMARY

The four steps to a quick but powerful assessment of drinking water quality are as follows:

1. Summarize water system configuration and operations parameters annually and compare to previous years.
2. Keep an updated timeline of water system changes and events.
3. Graph and update Lead and Copper Rule compliance data statistics over time to pinpoint time periods where water quality changes occurred and match those time periods to operational changes from Steps 1 and 2.

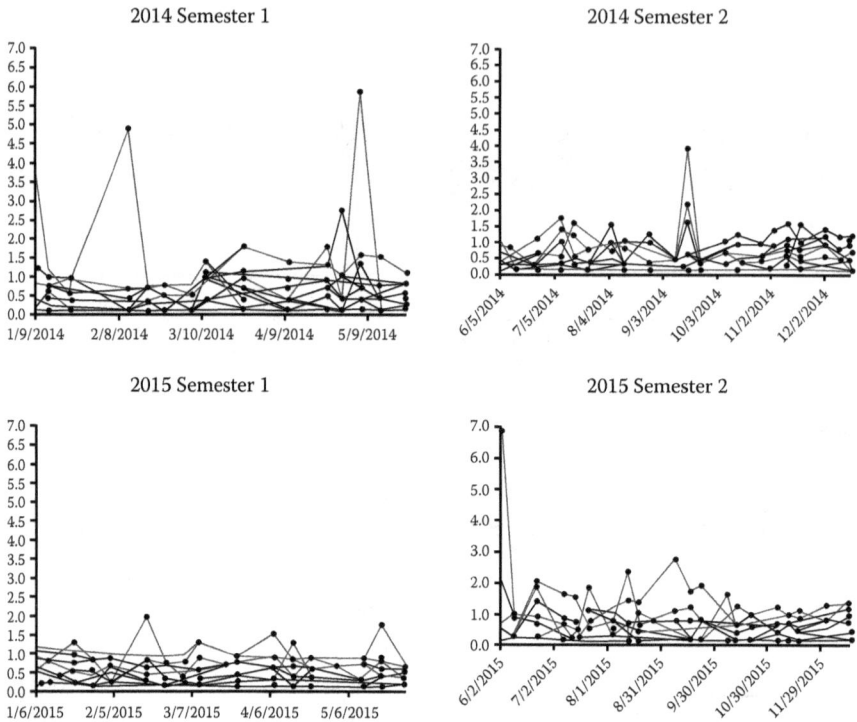

FIGURE 3.7 Comparison of turbidity in NTU at Total Coliform Rule sites over time.

4. Graph disinfection concentration and turbidity from each Total Coliform Rule distribution system site (or other routinely visited distribution system site) over time to pinpoint specific locations and times that disinfection is low and turbidity is high. Routinely evaluate and compare these parameters between time periods. Troubleshoot problem areas and change operations to keep disinfection at appropriate levels with low turbidity.

This exercise in using existing and easily accessible data in a water system as presented in this chapter can be sufficient for an ongoing program of water quality control and system improvement in many water systems, both small and large. If the directions in this chapter are carried out routinely, this book could end right here.

However, there are times when more information about water quality control is needed or desired. So, the book will continue for those readers who want to take further steps toward water quality control and system improvement.

4 Strategic Planning for Improved Water Quality

The method of water quality control described in Chapter 3 uses indicators of degraded water quality to indirectly identify water quality issues. Action can then be taken to bring the indicators back into a desired range. However, there are many times that one would want to know the specific reasons for degraded water quality in order to target these forces and control them.

In order to get more information about the influencing forces, we have to answer the questions posed in Chapter 1:

- Where do we measure the contaminant in a distribution system so that it is representative of what consumers are exposed to?
- How do we measure the contaminant if every building piping system is different?
- What do we measure to determine how the contaminant is created in the distribution system?
- How are we going to prevent or treat the contaminant throughout the distribution system?

These questions are answered in this chapter as a comprehensive distribution system sampling strategy is described.

4.1 SELECTION OF SAMPLING SITES

Monitoring in a distribution system to assess water quality has always been fraught with dilemmas. A good example of this is seen when assessing a water system's potential for transferring lead and copper from metal components into the drinking water. Lead and copper release into the water as well as other distribution system contaminants vary over location in the distribution system and over time. Water utility personnel cannot visit every faucet in every building to sample and certainly cannot do that routinely. Therefore, they must make decisions on where to draw the samples.

4.1.1 SAMPLE CRITICAL BUILDINGS

This is the reason that the United States Environmental Protection Agency (EPA) developed the strategy described for compliance sampling in the Lead and Copper Rule. Sampling sites are selected from residences most prone to having high lead levels, such as the ones with lead service lines and the ones with lead solder in copper piping. The number of sites for a utility is based on utility size. Compliance is based on comparing the 90th percentile concentration in the sample dataset to a regulatory Action Level that triggers action to modify the water or the system in order to lower lead and copper levels.

While this is a satisfactory strategy, there are issues with the technique. First of all, the technique may not capture the highest lead or copper levels in a building; only the first liter of water is taken from the kitchen sink, which does not capture water contained within a lead or copper water service line far upstream. Most importantly, it is very difficult for water utility personnel to enter privately owned buildings to obtain samples. Property owners tire of the sampling regime. For this reason, sampling can only be performed infrequently.

As data for analysis, lead and copper data from various buildings in a water system are challenging. It is difficult to justify comparing metal concentrations from endless variations of piping configuration, water usage, and water flow scenarios.

Even with these drawbacks, lead and copper data from critical sites in a distribution system can add to an understanding of lead and copper release and other water quality issues. For example, in Section 3.3, Lead and Copper Rule compliance data from critical residences were used to pinpoint time periods of major changes in water quality and correlate them with operational changes listed on the water system timeline.

Therefore, even with the issues described here, data from buildings in a distribution system that have been identified to be experiencing a particular water quality issue are important to study for clues to the origin of the issues.

4.1.2   SAMPLE A NETWORK OF SITES OVER THE DISTRIBUTION SYSTEM

A second strategy is to sample a network of sites geographically scattered around the distribution system. However, entering private property at routine intervals and working with a variety of sampling scenarios are still issues with this strategy.

But in Section 3.4 a network of sites was used for measuring two flowing water quality parameters (disinfection concentration and turbidity) as indirect indicators of water quality status. This made it possible to pinpoint areas of the distribution system that had higher potential for water quality issues to occur. It is very informative to track and take action on these patterns of water quality degradation over the network of locations and over time.

4.1.3   SAMPLE AT THE EXTREME SCENARIOS

Another strategy identifies the best and worst water quality scenarios in order to define the possible minimum and maximum contaminant concentrations in a water system. With this strategy, the water quality characteristics of a distribution system can be bracketed between two extremes—between the characteristics of the freshest water and those of the oldest water, that is, the water that has resided in the water system the longest time.

On one extreme, the freshest water with the highest concentrations of treatment chemicals can be found at the entry points to the distribution system. The effectiveness of the treatment chemicals is assumed to be at a maximum at these points. (Some water systems have only one entry point to the water distribution system;

this is common for a surface water system with one treatment facility. Other water systems, such as groundwater systems with multiple wells, can have many entry points.)

The opposite extreme of water quality at the entry points of a distribution system are locations of high water age in the distribution system. Water age is another term for residence time of water. High water age locations occur at the farthest reaches of a distribution system, at dead ends, and in areas of low water usage. As water age increases, both added and naturally occurring chemicals in the water have more time to interact chemically and microbiologically.

Other severe conditions in water distribution systems can also be good strategic choices to contrast with entry-point sites. Some of these sites occur where different water qualities blend together. This would be the case in a system with multiple water sources, such as in systems with multiple wells or in systems where groundwater supplements a surface water source. At these locations, water quality varies with multiple sources. The water quality swings can cause destabilizing conditions to existing chemical and microbiological accumulations on pipe walls. These swings, in turn, can impact corrosion, metal release, and other water quality issues.

Areas of high release of metals and other water quality issues in a distribution system may also be related to the age of the buildings in a neighborhood and the types of plumbing materials prevalent at the time the neighborhood was constructed. An inventory of system piping materials and age may provide better insight into selection of monitoring locations.

Customer complaints of discolored water, bad taste, or bad odor can additionally identify locations in the distribution system where corrosion of metals along with other water quality issues may be occurring. Such locations, especially if geographical patterns of complaints are evident, should also be considered when selecting critical sampling sites for routine monitoring.

4.1.4 COMBINING STRATEGIES

A combination of strategies, as advocated in this monitoring book, can be used for the routine tracking of distribution system contaminants.

The strategies are

- Capture of water quality at a subset of critical buildings in the distribution system
- Capture of water quality in a network of sampling sites around the distribution system
- Capture of water quality at the extremes of the distribution system

Sampling in buildings is described in Section 4.7. Capturing water quality in a network of sampling sites around the distribution system was described in Section 3.4. Capturing water quality at the extremes of the distribution system is described in Section 4.6.

But first, there are more fundamental concepts of sampling that need to be described.

4.2 STAGNATING VERSUS FLOWING WATER SAMPLES

After a sampling site selection strategy has been chosen, it must be determined whether flowing water or stagnating water should be captured for analysis. To understand when each type of sample is used, consider the brewing of beer as a frame of reference. To make beer, fresh ingredients with known characteristics are poured into a vat and then allowed to undergo chemical and microbiological changes for a period of time. At the end of the reaction period, the final product is withdrawn and analyzed to determine if the desired characteristics have been achieved. The brewing of beer is a batch process and the vat that holds the ingredients during the process is called a batch reactor.

Water pipes are also batch reactors. Fresh water flows into the pipe. Chemical and microbiological interactions occur during the time that the water is in the pipe. The water in the pipe can be withdrawn after a reaction period and analyzed to determine the outcome of the reactions. Examples of reaction outcomes in water are concentrations of lead, copper, and other metals or contaminants, final disinfectant concentration, formation of disinfection by-products, and population of microorganisms.

Consumers of water in a system do not typically receive water fresh from a well or treatment plant. They receive water after varying degrees of chemical and microbiological interactions have occurred in the piping of the water distribution system and in their buildings. Therefore, the batch reactor concept aids in studying the water quality that the consumer receives.

4.2.1 STAGNATING WATER SAMPLES

4.2.1.1 Buildings

Buildings connected to water systems are batch reactors. "Fresh" water flows in from a water main and then undergoes changes based on piping materials, residence time in pipes and tanks, flowrates, and quantity of water used. The Lead and Copper Rule uses residences with lead service lines or lead solder as the batch reactors in its sampling strategy. The Rule prescribes a minimum of six hours of reaction time, that is, six hours of water stagnating in the piping.

4.2.1.2 Pipe Loops

Given the problems already discussed with sampling from privately owned buildings, an apparatus was devised around 1990 to simulate a building piping system. It was called a "pipe loop apparatus," developed by the AWWA Research Foundation (AwwaRF, now renamed the Water Research Foundation) in anticipation of the 1991 Lead and Copper Rule (EES 1990). The original apparatus consisted of one or more lengths of lead or copper pipe, each pipe able to hold at least one liter of water so that a one-liter sample, similar to the Lead and Copper Rule residential compliance sample volume, could be obtained. The pipes were attached to a nonreactive plastic pipe that connected to "fresh" distribution system water. Flow meters controlled the flow to each pipe. Water flowed from the distribution system, through the pipes and to waste, as occurs in building plumbing. The water flowing from the distribution system was controlled by a programmable timer opening and closing a valve to turn water on and off, similar to the water usage pattern in a building.

The timer also allowed for a period of stagnation as described for reaction time in batch reactors. Because the Lead and Copper Rule calls for a minimum of six hours stagnation in residential plumbing before sampling, six hours became a common stagnation time used before sampling a pipe loop apparatus.

The original AwwaRF pipe loop apparatus was only used for chemical comparison tests of corrosion-control products in entry-point water; it was not used for routine monitoring of water quality at various locations around a distribution system. But, due to its size, even if it would have been used for distribution system monitoring, it would have been very awkward to place out in the water system.

4.2.1.3 Mini Pipe Loops

Mini pipe loops were developed to utilize the AwwaRF pipe loop concept as a means to monitor around a water distribution system (Cantor et al. 2000). They were originally constructed in 1996 at one-quarter the size of a standard pipe loop and skid-mounted. While the full-size apparatus might take up a large area at a water treatment plant, the mini pipe loops were portable to a roughly three-foot-by-five-foot floor space in the distribution system. Copper mini pipe loops were first used to place around a distribution system to monitor system water for effects of an added chemical (polyphosphate) that was thought to be increasing the copper levels in the water (Cantor et al. 2000). See Figure 4.1. This strategy successfully identified the chemical as the major influence of the copper problem and also identified when residences could be resampled to show lower copper concentrations after removal of the chemical.

In 1999, mini pipe loops were constructed for a research project to test the effects of chlorine addition on corrosion (Cantor, Park, Vaiyavatjamai 2003). See Figure 4.2. Three metals and three chemical treatments were studied using nine pipe loops. The complete apparatus fit into the corner of a small well house.

4.2.1.4 PRS Monitoring Stations

After the experiences with the mini pipe loops, it became obvious that a distribution system monitoring station used as a routine gauge of water quality adds much insight into lead and copper release and other distribution system water quality issues during routine water system operations and during system operational changes. Based on these experiences, criteria for an ideal distribution system monitoring station were developed. The ideal monitoring station should

- Be relatively inexpensive
- Be easy to move from location to location
- Not take up excessive space
- Be cushioned and isolated from vibrations that can interfere with pipe accumulation release
- Be straightforward to operate
- Be used for either off-line chemical treatment comparison tests or as a gauge of water quality around a distribution system

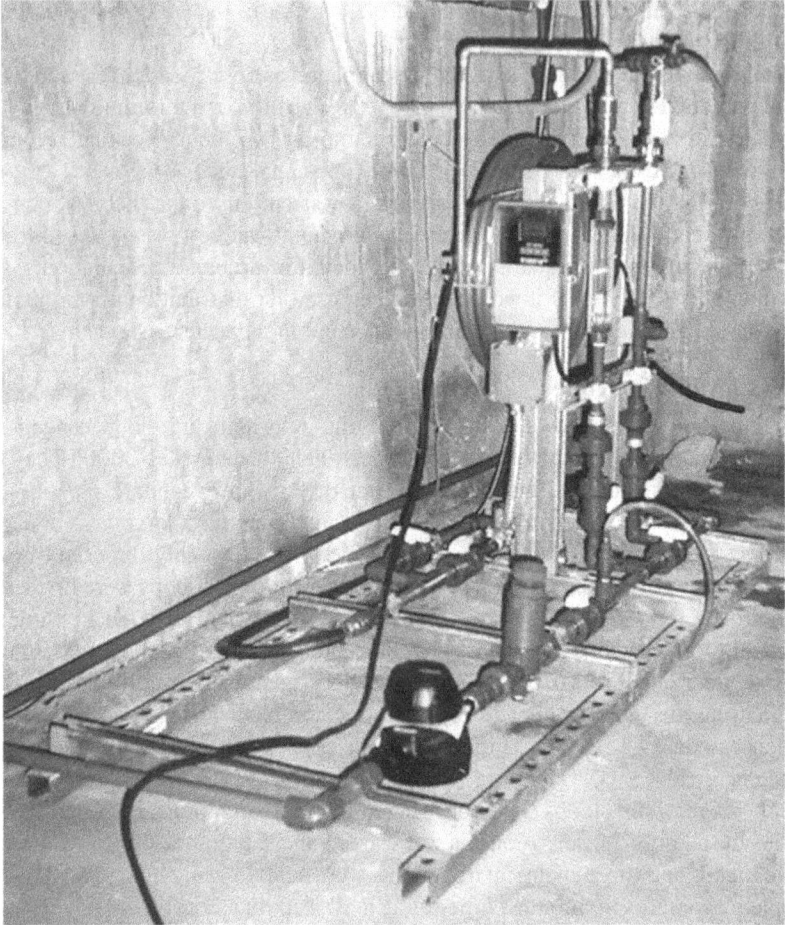

FIGURE 4.1   Mini pipe loop used as a distribution system monitoring station in 1996.

- Allow for routine collection of water samples
- Be easily accessible to water utility staff
- Have a uniform configuration
- Have uniform water flow rate and water usage
- Have a steady pressure
- Be capable of water stagnation for a uniform time
- Capture routine operation of the water system as well as treatment and operational changes that may occur over time
- Capture the interaction of pipe scales and films with the water
- Make pipe film and scale analysis possible and economical so that the young scales developed during the monitoring period can be compared to older existing scales on actual water system pipe
- Not destroy pipe film and scale during the metal sample preparation

FIGURE 4.2   Mini pipe loops of three metals for chemical treatment comparison in 1999.

To meet these criteria, the mini pipe loops were modified, in 2006, into the Process Research Solutions (PRS) Monitoring Station (Cantor 2008, 2009). See Figure 4.3. Instead of a piece of coiled pipe, stacks of metal plates that are secured in a section of larger plastic pipe are exposed to system water. The surface area of the metal plates to the volume of water held in the test chamber is similar to a 1.77-inch diameter pipe used as a pipe loop. This equivalent-diameter pipe was the most economically and physically practical to achieve. See Figure 4.4 to view the metal plates stacked inside the test chambers before the chambers are sealed.

FIGURE 4.3 PRS Monitoring Station, a standardized distribution system monitoring station.

FIGURE 4.4 View of plates set inside and stacked in test chambers in a PRS Monitoring Station before the test chambers are closed.

The PRS Monitoring Station is operated in the same way that standard or mini pipe loops are operated. A timer controls the flow of water through the device as well as the time of stagnation. First-draw water samples are taken from the test chambers at the end of a prescribed stagnation period, collecting water that has been exposed to the metal surfaces. See Appendix A for more details of assembling, installing, and operating a PRS Monitoring Station.

4.2.2 FLOWING WATER SAMPLES

As in the beer-making analogy, the characteristics of the water flowing into a batch reactor are also desired to characterize the "fresh ingredients." When sampling in buildings, water should be allowed to run in order to fill the pipes with "fresh" water from the water main. However, in order to obtain a sample that is representative of typical water for consumption, the flow rate of the water from the sample tap should be similar to filling a glass of water for drinking. Too high a flow might scour excess pipe wall debris into the sample; too low a flow might eliminate typical contributions from pipe wall accumulations.

With the PRS Monitoring Station test chambers acting as the "batch reactors," the "fresh" water samples are obtained from the station's influent sample tap while water is flowing.

Also, in order to contrast the high water age flowing water, flowing water samples from the entry points to the distribution systems should also be obtained at the same frequency to characterize the "freshest" water in the system.

4.3   METAL SURFACE DILEMMA

In water distribution systems, all piping, whether in the distribution system or in buildings, has varying degrees of existing chemical scales and biofilms that have built up over time. This is because all water is a complex solution of many naturally occurring and added chemicals and a naturally occurring potpourri of soil and air-borne microorganisms. Chemical scales precipitate onto the pipe walls over time; they are intertwined by biofilm development from microorganisms as the water environment allows. It is the interaction of the pipe wall accumulations with the adjacent water that shapes water quality.

To represent actual system pipes, the original AwwaRF pipe loop apparatus was intended to hold old lead pipes harvested from the distribution system and carrying decades of chemical scales and biofilms on their surfaces. This is good in theory; however, experimental dilemmas arose. When old pipes harvested from water distribution systems were used, there was no control over how representative each harvested pipe may or may not have been in the distribution system. In addition, metal particulates from the debris on the old pipe walls could be disturbed and interfere with the metal concentration data for a long period of time (AwwaRF and DVGW-TZW 1996).

Having new metal surfaces is equally problematic. Lead and copper transfer from bare metal is not representative of metal transfer in the water system. It takes several weeks to months for metal oxide and carbonate scales to form and cover the bare metal (Section 2.2.1). The metal release then becomes more typical of the water system but still is not the same in response as the old water system scales.

The PRS Monitoring Station begins with new metal surfaces on the internal metal plates. The metal release from the plates is often high at first. To eliminate these unrepresentative data, monitoring of metal release in the test chambers is typically not begun until a month after the station has been operating. Then, the data are studied to determine when a steady range of metal release is achieved. Every water system is different as to how long it takes for lead and copper concentrations to stabilize in the water.

This is an experimental design choice to start a monitoring period with known surface characteristics (bare metal), as opposed to selecting random pipes from a distribution system with existing accumulations of unknown quantity and composition. With the choice of bare metal surfaces, the monitoring apparatus configuration can be reproduced from monitoring station to monitoring station and from water system to water system.

In addition, the metal plates of the PRS Monitoring Station have an advantage over new metal in a pipe loop apparatus. That is, all the metal surfaces are in a configuration of 2-inch-by-2-inch-by-1/16-inch-thick metal plates. These metal squares are removed from the test chambers at the end of a monitoring project period and sent for both microbiological and chemical analysis. These are the same types of analyses performed on pipes harvested from distribution systems, but harvested pipes must be cut open in order to study their accumulations. The cutting operation can contaminate the surfaces to be studied. With the PRS Monitoring Station metal plates, they need little further processing before study. In this way, test chamber metal plates are studied and the chemical scales and biofilms are characterized.

The metal plate scales have not had as much time to form as the scales on existing piping. When there have been opportunities to compare them with scales on existing piping, there are similarities but some of the compounds are farther away from their thermodynamically stable forms. The younger scales show the direction in which the scales are heading. They also show similar extraneous elements and minerals as existing piping scales, as well as showing how phosphorus, when dosed into the water, is interacting with lead and copper. That is, the early interactions that form pipe wall accumulations can be studied and influencing forces possibly identified. This also sheds more light on the water quality data obtained during the operation of the PRS Monitoring Station.

The metal plate information should be compared to analyses of one or more pipes of the same material harvested from the distribution system and to water quality data obtained in sampling critical buildings.

See Appendix B for details of the metal plate and harvested pipe chemical and microbiological analyses.

4.4 SELECTION OF WATER QUALITY PARAMETERS

There can be many factors at work alone or simultaneously that can shape distribution system water quality and possibly create undesirable characteristics. Not knowing what influencing factors are occurring in any given water system, quite a number of water quality parameters can be measured as guided by the comprehensive perspective of water quality described in Chapter 2.

To organize the many possibilities of measurements that can be made, water quality parameters can be sorted into the three main categories of influencing factors described in Chapter 2. Groups of water quality parameters to measure in water samples are as follows:

- Parameters involved in uniform corrosion of metals
- Parameters involved in the biostability of water
- Parameters involved in pipe wall scale formation or dissolution

Note that several water quality parameters can fall into more than one of these categories.

These categories of influencing factors and parameters were not arbitrarily chosen. They were selected after studying accumulations on PRS Monitoring Station test chamber metal plates and also components in the test chamber stagnating water and summarizing observations over many years of working with the monitoring stations. There is room to add any number of water quality parameters to these categories; using the categories to group water quality parameters is an organizational method, but one based on many observations.

Regarding uniform corrosion of metals, several types of chemistry should be considered. Carbonate chemistry is the focus of the Lead and Copper Rule and must be studied in each water system. Alkalinity, pH, temperature, conductivity (total dissolved solids), and hardness all describe carbonate chemistry. Chloride and sulfate, common chemicals in drinking water, also appear to affect the dissolution of metals. The chemistry of higher oxidation states must also be considered, as discussed in Section 2.2.1, to look for the potential to form insoluble lead oxide compounds. Therefore, the oxidation–reduction potential (ORP) of the water is important to track. Phosphate chemistry is important because the Lead and Copper Rule relies heavily on the addition of phosphate to control lead and copper. Typical phosphate corrosion control products include both orthophosphate and polyphosphate fractions. Total phosphorus and orthophosphate can be measured and polyphosphate found by subtraction.

Regarding biostability, there is an accessible and affordable analysis available to track the population size of all microorganisms (with the exception of viruses) in a water system. That is the analysis for the energy molecule of living cells, adenosine triphosphate or ATP. If ATP is measured in a water sample, there are living organisms in the water. The population size can be assessed in the context of concentrations of nutrients for encouraging growth of microorganisms—nitrogen, phosphorus, and carbon compounds. The population size can also be assessed in the context of disinfection concentration for eliminating microorganisms. Disinfection concentration can be fractionated into active disinfection and disinfection chemical combined into other compounds.

Regarding scale formation or dissolution, a number of metals can be found in drinking water. Some scale-forming metals can adsorb dissolved metals that pass the scale in minute quantities in the adjacent water. The dissolved metals accumulate on the surfaces of the metal scales over time. Eventually, the pipe wall scales crumble and the adsorbed metals are transported through the water system attached to scale particulates to consumers' taps at higher concentrations than normally experienced. Iron, manganese, and aluminum are known for this (Schock et al. 2014).

There are several plumbing-related metals that can corrode in water systems: lead, copper, cadmium, chromium, cobalt, nickel, tin, and zinc.

There are minerals related to the natural hardness or softening treatment of the water: calcium, magnesium, strontium, barium, potassium, and sodium.

There are also other metals that occur naturally in the water that may be of interest to track, such as arsenic.

A panel of metals can be obtained from laboratories by means of a metal scan from an inductively coupled plasma (ICP) analysis.

A measurement of any metal concentration can be further fractionated into dissolved and particulate forms of that metal. To do this, a portion of a water sample must be filtered through a 0.45-μm filter (APHA et al. 1995). The metals analyzed in the unfiltered sample portion represent total metal concentrations. The metals analyzed in the filtered sample portion represent dissolved metal concentrations. Particulate metals equal total metals minus dissolved metal concentrations. Ideally, water samples should be field-filtered immediately after obtaining the samples so that metals do not change from dissolved to particulate forms or vice versa or adhere to sample bottle walls (Cantor 2006). If this cannot occur, then lab filtration will have to suffice but with the suspicion that metals may have changed form.

General water quality parameters describing scale formation or dissolution are turbidity (the measurement of particulate solids entrained in the water) and conductivity (the measurement of dissolved solids in the water).

Table 4.1 summarizes water quality parameters sorted into the three organizational categories of factors that shape distribution system water quality. See Appendix C for more details about the measurement of each water quality parameter. Other water quality parameters can be added to the groups listed in Table 4.1 based on the specific requirements of an individual water system.

TABLE 4.1
Water Quality Parameters Related to Categories of Factors That Shape Distribution System Water Quality

Uniform Corrosion of Metals	Biostability	Scale Formation and Dissolution
Total alkalinity	Adenosine triphosphate (ATP) as a measure of microbiological population	Turbidity
pH	Free chlorine or monochloramine as the active disinfectant	Conductivity
Temperature	Total chlorine including the active disinfectant plus chlorine combined in other compounds	Total, dissolved, and particulate concentrations of the following metals:
Conductivity	Dissolved organic carbon	Lead, copper
Total hardness	Ammonia nitrogen	Iron, manganese, aluminum
Chloride	Nitrite nitrogen	Cadmium, chromium, cobalt
Sulfate	Nitrate nitrogen	Nickel, tin, zinc
Total phosphorus	Total phosphorus	Calcium, magnesium
Orthophosphate	Total alkalinity	Barium, strontium
Oxidation/reduction potential (ORP)	pH	Sodium, potassium
	Temperature	Arsenic
	Oxidation/reduction potential (ORP)	

4.5   FREQUENCY OF WATER QUALITY PARAMETER SAMPLING

The frequency of sampling should be based on the nature of a parameter's variation in a given water system. It is necessary to sample more frequently at first until the variation of a parameter is understood and an adequate frequency can be set.

Budget and labor availability for monitoring may force compromises in data-gathering frequency and number of parameters.

4.6   MONITORING PLAN FOR WATER DISTRIBUTION SYSTEMS

Based on past experiences with PRS Monitoring Stations in a variety of water systems, a distribution system monitoring plan is suggested here. The plan can be used as a starting point and modified to the needs of an individual water system.

For sampling site selection, the combination strategy discussed in Section 4.1 is recommended. This includes sampling a network of sites over the distribution system. In order to carry this out, it is hoped that water utilities adopt Chapter 3's ongoing assessment on water system configuration and operations parameters, timelines, Lead and Copper Rule data, and Total Coliform Rule site parameters as a routine practice.

PRS Monitoring Stations can be used to add the most detailed information of the combination site selection strategy—sampling at extreme scenarios. (When PRS Monitoring Stations are mentioned in this discussion, other apparatuses similar to AwwaRF pipe loops can be used instead.) That is, PRS Monitoring Stations can be placed in the distribution system at locations of high water age. Common installation sites are booster pumping stations. Monitoring stations can also be located in any building that water utility personnel have routine access to, such as a fire station,

---

TABLE 4.2
Sampling Sites for a Typical Distribution System Monitoring Plan

Abbreviation	Description
EP-X	Flowing water at entry points to the distribution system where X is the identification number or name of the entry-point location.
MS-Y_Inf	Flowing water at the influent sample tap of the PRS Monitoring Station where Y is the identification number or name of the monitoring station; this site represents flowing system water characteristics at the influent to the monitoring station.
MS-Y_[Metal]-Z	Stagnating water in each test chamber of the PRS Monitoring Station where Y is the identification number or name of the monitoring station. The type of metal used in the test chamber should be designated. (Example: [Metal] = Pb for a test chamber of lead plates.) Z is the identification number of the test chamber if there are duplicate test chambers with the same metal. For example, if there are duplicate lead test chambers on Monitoring Station 1, they can be designated as MS-1_Pb-1 and MS-1_Pb-2.

---

a city government building, or a school. Even in a large water system, installing only one monitoring station at one high water age location is very informative. Planning should assess whether more critical locations in the distribution system should be monitored and whether a monitoring station should be located at one or more entry points to the distribution system. A typical monitoring plan includes performing flowing water sampling at all entry points to the distribution system plus flowing and stagnating water sampling at one PRS Monitoring Station installed in a high water age location. Sampling locations can be designated, as shown in Table 4.2.

The complete monitoring plan is shown in Table 4.3. The plan can be altered to better fit the needs of the water system and the budget. However, it is important to incorporate parameters from the biostability and chemical scale groups of factors and not just the uniform corrosion factors. Also, it is important to measure metal release and other relevant stagnating water results more than once a month. Otherwise, important changes in concentrations may be missed.

TABLE 4.3
Typical Distribution System Monitoring Plan

TestType ⟶	Flowing Field Tests	Flowing Lab Tests	Stagnating Lab Tests
Sampling Sites* ⟶	EP-X, MS-Y_Inf	MS-Y_Inf	MS-Y_[Metal]-Z
Frequency			
Weekly	Flow meter totalizer readings, pH, temperature, ORP, turbidity, conductivity, total chlorine, free chlorine or monochloramine, orthophosphate or silicate, if relevant		
Biweekly		Total metal scan, ATP	Total metal scan, dissolved metal scan, ATP
Monthly		Total alkalinity, chloride, sulfate, dissolved organic carbon, total phosphorus, ammonia, nitrite, nitrate	

* See Table 4.2 for sampling site designations.

Allowing a monitoring program with the PRS Monitoring Stations to proceed for at least a year gives insight into water quality changes that are temperature-dependent and manifest themselves seasonally.

At the end of a monitoring period, the PRS Monitoring Stations are shut down and the metal plates are carefully removed and sent to microbiological and chemical laboratories for analysis of layers of metal surface accumulations. This gives insight into many of the chemical and microbiological mechanisms that are shaping the water quality and can provide commentary on observations in the water analysis data. Additional harvesting of one or more distribution system water service lines and analysis of existing pipe wall accumulations can also increase insight into influencing factors of water quality and results can be compared to the PRS Monitoring Station metal plates.

4.7   MONITORING PLAN FOR BUILDING PLUMBING

There is also a third component to the combination sampling strategy; sampling of critical buildings in the distribution system can identify possible localized issues but can also verify water quality trends. In addition to the PRS Monitoring Station and entry-point monitoring, one or more critical buildings should also be sampled. Critical buildings are described earlier, in Section 4.1.1.

There are several strategies that can be used for sampling of buildings. Building monitoring strategies can be evaluated based on budget and access to desired sites.

One strategy is to locate a faucet in a critical building that can be routinely accessed by water utility personnel. In one project, owners of a commercial building allowed water utility personnel to enter the building early in the morning as often as desired to gather stagnating water samples from a particular faucet before occupants of the building arrived. Then, the sampler allowed the water to flow until cooler in temperature in order to obtain water samples of "fresher" water. The same analyses were performed on the stagnating and flowing water samples as were performed on the PRS Monitoring Station samples. The building faucet was visited once a week, the same frequency as the PRS Monitoring Station sampling. This was similar to Lead and Copper Rule compliance sampling protocols, but with a number of additional steps.

In other PRS Monitoring Station projects, critical residential buildings have been selected for sampling every three to six months during the monitoring station operation time period. The residential buildings can be sampled using protocols similar to the Lead and Copper Rule sampling or they can be subjected to profile sampling.

Profile sampling is a technique typically for lead and copper release assessment where sequential liters of stagnating water are drawn from a building's plumbing system. Analysis of each liter sample can describe where lead or copper reach higher concentrations relative to other locations in the plumbing system. That location is related to piping material, giving insight into the piping materials and locations in an individual building where lead or copper can be problematic. Lead and copper concentrations found by this method quite often show that the first-draw liter of water, as prescribed to be taken for Lead and Copper Rule compliance, does not necessarily represent the maximum lead or copper concentration that could potentially reach the water consumers in that building (Cornwell and Brown 2015). The protocols can be modified in order to include more water quality analyses of samples as performed at the stagnating and flowing water of the PRS Monitoring Station. A flowing water sample can also be obtained.

The following sampling protocol can be used or modified for profile sampling of residences:

- Select houses with a service line of the material similar to one of the PRS Monitoring Station's test chambers or any material of interest.
- Using a hardness test strip, confirm that the kitchen cold water tap is not softened.
- Estimate the volume of water from the kitchen tap through the service line and determine the number of liter-bottles required from sample tap to water main. Add two more liter-bottles to obtain water samples from within the water main.
- Run water at "normal" flow rate at the kitchen tap until the water is cold—about five minutes. This helps to ensure that all sampling taps in all buildings began the experiment with "fresh" water from the water main. "Normal" flow rate is like filling a glass of water; it is not a flushing velocity, nor is it a trickle of flow.
- If possible, take flowing water samples for flowing field tests and lab tests as listed in Table 4.3. Use a "normal" flow rate to capture the sample in field and laboratory analysis sample bottles.
- Check for automatic water use, such as ice makers, softeners, and humidifiers, and turn off.
- Stagnate the water six hours minimum or overnight.
- Keep the faucet aerator on as is expected in Lead and Copper Rule compliance sampling.
- Using "normal" flow rate, take samples to reach from tap through lead service line into water main in separate one-liter bottles. Wide-mouth bottles are to be used so that water flow into sample bottles can be similar to filling a glass of water. There should be no acid in any sample bottle in case a field or laboratory filtration of a portion of the sample must be performed. With the filtration, separate dissolved and total metal concentrations can be captured before acid preservation of the sample dissolves all metals. (This is explained in Appendix C.)
- After stagnation sampling, if flowing water samples have not been taken, run water until it is cold and take the flowing water samples now.
- Turn on automatic water-use appliances before leaving the house.
- Send all bottles to the laboratory for analysis. Stagnation samples are to be analyzed for stagnating water quality parameters listed in Table 4.3, as budget allows.

The sampling protocols described in this section are for gathering drinking water samples in buildings either for a general assessment of drinking water quality in a distribution system or for comparison to a more detailed assessment of distribution system drinking water quality from a PRS Monitoring Station. For a detailed investigation of water quality issues in a specific building, all interior piping systems must be assessed—cold potable water, hot water, softened water, and other treated water. A description of a building water quality investigation is out of the scope of this book.

4.8   MONITORING: THE PATH TO CONTROL OF WATER QUALITY

Monitoring and performing measurements in any system has always been controversial. The effort to measure and study results takes time and money away from the direct operations of the system. Dr. W. Edwards Deming, a champion of industrial process improvement, found this to be a barrier as the United States entered World War II. He promoted the idea of routine measurement and assessment of industrial processes so that the manufacture of war materials would produce high-quality products to support the military personnel who depended on them (Wheeler and Chambers 1992). Engineers and technicians understood the value of minimizing waste from discarded products that were not to required specifications and of increasing product dependability by spending resources on routine measurement and assessment. However, corporate management could not understand these intangible benefits of preventing consumers' problems from occurring; the prevention of a problem cannot always be proved or stated financially. Dr. Deming's protocols of process control and improvement were not adopted by postwar corporate America. Ironically, Dr. Deming was sent to Japan by the U.S. government after the war to help reestablish the Japanese economy. Japanese industrialists, both the technical personnel and the management, did listen to Dr. Deming's message (Wheeler and Chambers 1992). This is the reason that Japanese electronics and cars exceeded the quality of American products for many years after World War II.

Just as Dr. Deming did for manufactured products, this book promotes techniques for routine comprehensive measurement of water quality characteristics for:

- Definition of influencing forces on water quality in individual water systems
- Guidance on controlling undesired water quality characteristics
- Ongoing control of water quality and water system operations improvement

Figure 4.5 displays the path to control of drinking water quality. The path starts with actions that were described in Chapter 3—gathering existing water system information and performing relatively simple analyses with the data in order to identify water quality issues and goals to strive for. Actions continue as described in this chapter in developing a monitoring plan for the distribution system. A monitoring plan should be developed and carried out even if no water quality issues were identified. This is because a water quality issue may not be recognizable by current assumptions regarding water quality but can become evident when the water quality has been studied in the comprehensive manner described here. If a water quality issue cannot be found, then the monitoring program can proceed to routine monitoring and system improvement where steps can be made to tighten water quality specifications and optimize water system operations.

Gather existing information describing the water system.
See Chapter 3 for details.

Perform analyses described in Chapter 3.

From the Chapter 3 assessment, identify water quality issues and
goals to strive for.

Develop a comprehensive and representative monitoring plan as
described in Chapter 4, even if there are no identified water
quality issues.

Begin initial monitoring.

Assess monitoring data using data analytical techniques in
Chapter 5. Also use known scientific concepts if the concepts are
confirmed for the individual water system with the monitoring
data relationships and correlations assessment.

Yes — All water quality goals met? — No — Form hypotheses as to which water quality
parameters are the influencing factors in shaping
certain water quality characteristics.

Yes — All water quality goals met? — No — Incrementally change selected water quality
parameters.

Yes

Moving toward desired water quality goals? — Assess the additional monitoring data.

Continue monitoring.

No — Re-form hypotheses — Yes

Perform routine monitoring to check that
each water quality goal continues to be met. — Water system changes made?

No

Narrow variation, move the average
toward an optimum level or both
for process improvement (Deming
1993). — No — Moving away from desired water quality goals?

Take immediate action to find and "eliminate
the special cause" of the water quality
change (Deming 1993). — Yes

Strategic planning

Initial monitoring

Hypothesis testing and remediation

Routine monitoring and system improvement

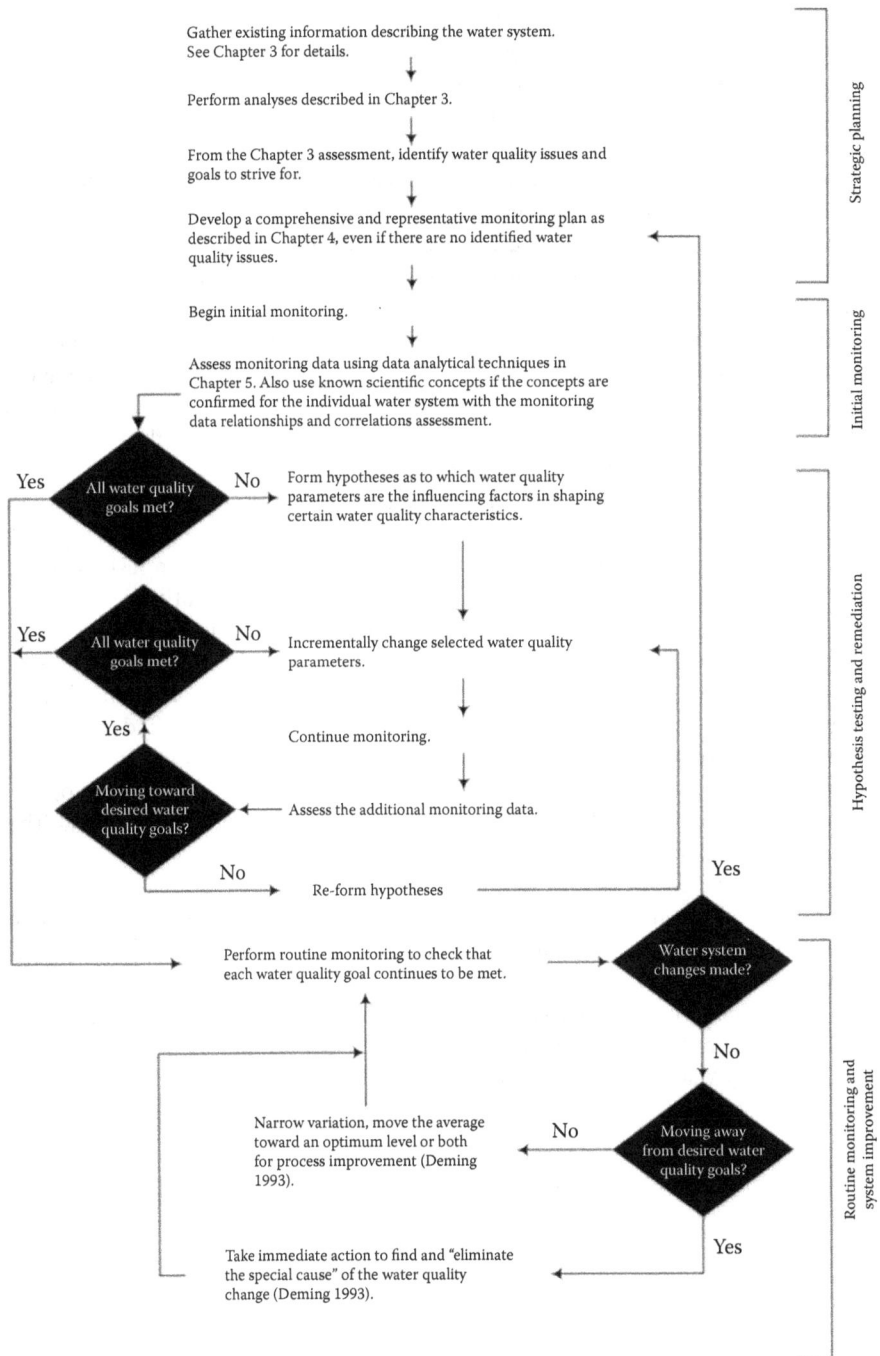

FIGURE 4.5 Path to control of drinking water quality.

With the activities in Figure 4.5, a large quantity of data is collected and must be analyzed. Techniques of data analysis will be discussed in Chapter 5.

Armed with monitoring data, hypotheses for the shaping of water quality can be formed. It is a big step to go from a lot of water quality monitoring data to defining the influencing forces on water quality in an individual water system. The monitoring is performed on a dynamic system undergoing multiple simultaneous influences. It is not performed on a highly controlled and isolated experiment. Therefore, care must be taken in not assuming cause and effect between water quality parameters. The best that can be done is to keep an open mind and to form hypotheses. Then, the hypotheses must be tested.

To test a hypothesis, one or more water quality parameters can be changed incrementally and slowly so as to not cause an upset of the chemical and microbiological accumulations on the pipe walls. Monitoring can continue to determine if there are trends toward the desired water quality characteristics or away from them. Data can also be studied at the same time to determine if there are negative side effects on other water quality characteristics. If the system is moving toward desired water quality characteristics, then the next incremental change in the selected water quality parameters can be made. Hypothesis testing continues in an iterative fashion along with monitoring until water quality goals are reached, as shown in Figure 4.5.

When all water quality goals have been met and all water quality characteristics are at desired levels, the monitoring program then moves into an ongoing process. If water system changes are made, then there may be a need to revise the monitoring program and begin again. Otherwise, the ongoing data analysis may find trends that the system is moving away from desired water quality characteristics. In that case, action should be taken immediately to determine what aspect of the system operations changes might have pushed the water quality in an undesired direction. The influencing factor should be eliminated and testing continued to determine when the trends are as desired again.

If the water system stays within the desired characteristics' ranges, then this is an opportunity to set more stringent goals—to improve average values of water quality parameters and to hold those parameters within a narrower range of variation. This is the quality control and system improvement that Dr. Deming advocated for manufactured products. It is possible for drinking water utilities to achieve this state of high-quality water and water system operation enlightenment.

5 Data Management and Analysis

This book so far has described methods by which all water system operations decisions are based on water quality data. Review Figure 4.5 to see that, with these methods, water quality problems are described by monitoring data. The solutions to water quality problems are theorized by studying the water quality data. The theories are tested and water quality improved by making small system changes, gathering more data, and reassessing if goals are being met. Collection of data continues in water systems with no water quality issues in order to catch early trends of water quality degradation and take action to correct the trends. This is empirical science. This is Dr. Deming's quality control and process improvement.

And, let's face it, this is a lot of data to work with. If there is not a way to keep the data organized and to easily understand what the data are "saying," then the data will sit on sheets of paper or take up space on computer hard drives and not be used, making the water quality monitoring efforts a waste of time and money. In other words, if it becomes difficult to work with the data, the data most likely will not be used.

This chapter describes ways to work with the water quality monitoring data and make them useful—more than that, make them invaluable—to water system operations.

5.1 THE SANCTITY OF DATA

Data management begins with respecting data. This sounds crazy. But think about the effort it takes to obtain one Lead and Copper Rule compliance sample. Arrangements must be made with the owners of a residence. The owners have to agree to allow the water in their house to sit stagnant for a minimum of six hours, and not flush toilets or run water for drinking or bathing. Then, they have to fill a bottle with the first stagnating water out of the kitchen faucet. They have to make arrangements for water utility personnel to pick up the bottle. The water utility personnel have to deliver the bottle to a laboratory, and on and on, until the final data point resides on the regulatory agency's computer. That one data point represents much time and effort and that deserves respect.

Another reason to respect data is to be able to reconstruct the circumstances under which the data were taken. In basic chemistry class, students are taught to keep their laboratory notes in a bound notebook and write in ink. If mistakes are made, the mistake is to be crossed out and the correction written next to the original entry. The mistake may be later found to be the correct information after all and that can be retrieved. Fast forward from basic chemistry class to an engineer, scientist, or technician collecting data in the field. After a long day visiting faucets or hydrants around a city and running field tests on each sample, the information written on the field sheets—crossed-out mistakes, correct measurements, duplicate measurements, comments about interferences to the field work—is the only way to answer any question about a confusing result. That is, data that are given respect can "answer" any future questions about them.

One more reason to respect data is to provide transparency to water system operations. If a person is not able to show up to work suddenly, properly written and stored data can communicate their work to a coworker to carry on for them. If consumers have a concern about an issue with the water system operation, the data can communicate the efforts made to deal with the issue. That is, transparency is important to a coordinated effort in operating the water system and in consumer confidence in the operations of the system.

In order to respect data, data should never be altered without a documented trail of written reasons to justify an alteration. It is a matter of good science, good water system operation, and good ethics.

5.2   DATA STORAGE AND RETRIEVAL

Data should be stored electronically with proper backup. Without the capabilities that computers provide, large quantities of data cannot be utilized.

Paper field sheets and laboratory reports should be scanned and stored as files on a computer. If there is any question in the future, an original data point can be located to determine if it has been changed or corrupted.

Data should be entered into a computer data-handling program from the field sheets and laboratory reports. If data are stored in the computer properly, specific information can be retrieved quickly and can be studied. The proper way to store data in a computer is in a relational database such as Microsoft Office Access® (Microsoft Corporation, Redmond, WA). A relational database is designed to allow for quick and efficient retrieval of data. It also can have safeguards in place so that data input is accurate and consistent.

It is best not to use a spreadsheet to store data! Spreadsheets, such as the commonly used application Microsoft Office Excel® (Microsoft Corporation, Redmond, WA), are for data analysis and reporting and do not have the important data storage features of a relational database. It is the existence of relational databases that allows us to collect and study data to a degree that was not possible in the past. However, there is not always the ability to store data in a database at a water utility. In this case, spreadsheets can be used with precautions taken. Keep one spreadsheet that only holds the data. If data analyses are to be made, a copy should be made of the original spreadsheet and analyses performed in the copy.

For ideal data management:

- Store monitoring data in a relational database file. As a last resort, data can be stored in a spreadsheet marked for data storage only, no analyses.
- Assure that the database or spreadsheet data file is backed up properly.
- Have safeguards in place for accurate and consistent data input, being consistent in naming and spelling sampling sites, water quality parameters, and units of measurement.

- Be able to retrieve specific data quickly.
- Transfer a copy of specific data from the database or spreadsheet data file to a separate spreadsheet for study.

For data analysis, basic spreadsheet skills are necessary—copying and pasting data columns, graphing data points, and calculating basic statistics. Ability to use special data analysis functions in the spreadsheet is also desired. Have either in-house expertise or outsourced help available.

5.3 DATA DISTRIBUTIONS

There are many ways to analyze a set of data. Each different method allows a person to think about the dataset from different perspectives. One early step in analyzing data is to see how the data are distributed between the maximum and minimum values of the dataset. Table 5.1 shows two sets of lead concentrations from two different periods of Lead and Copper Rule compliance sampling.

Statistics are calculated for each dataset in Table 5.1. Spreadsheet applications typically have built-in functions to calculate statistics. The functions in Microsoft Office Excel are as follows:

- =Count([data]) is the number of data points in the dataset.
- =Min([data]) is the minimum value in the dataset.

TABLE 5.1
Example of Statistics for Two Datasets from Lead and Copper Rule Compliance Sampling

	Dataset 1	Dataset 2
	Lead in μg/L	Lead in μg/L
	1	1
	2	2
	3	3
	4	4
	5	5
	6	6
	7	7
	8	8
	9	9
	10	50
Count	10	10
Minimum	1	1
Median	5.5	5.5
Average	5.5	9.5
90th percentile	9.1	13.1
Maximum	10	50

- =Median([data]) is the median or 50th percentile of the dataset where 50% of the values are above the value calculated and 50% are below.
- =Average([data]) is the average of the dataset where the values are summed and then divided by the number of data points.
- =Percentile([data],0.9) is the 90th percentile of the dataset where 10% of the values are above the value calculated and 90% are below.
- =Max([data]) is the maximum value of the dataset.

It can be seen from Table 5.1 that the two datasets have the same median value but different averages because of a higher maximum in Dataset 2. The 90th percentile is also higher in Dataset 2.

In order to study how the data are distributed between the minimum and maximum values in each dataset, histograms are used. Spreadsheet applications typically have a histogram function. A histogram counts how many data-point values fall into certain ranges (also called "bins"). Table 5.2 shows the count (also called "frequency") of data points in each range.

The number of data points in each range can be expressed as the cumulative number of data points that are equal to or less than a given range. Table 5.2 shows that 100% of Dataset 1's values are less than 10 while only 90% of Dataset 2's values are less than 10.

Histograms can be plotted on graphs for a better visual understanding of the data. The frequency of data points in each range is shown in Figure 5.1. The percent cumulative number of data points in each range is shown in Figure 5.2.

There is yet another way to show the distribution of data in a dataset and that is by using box and whisker plots. In box and whisker plots, data points are divided into four groups (quartiles), as shown in Figure 5.3. The grey box in Figure 5.3

TABLE 5.2

Histogram Information of the Datasets in Table 5.1

Ranges of Lead in µg/L	Frequency or Number of Data Points in Range	% Cumulative Number of Data Points in Range
	Dataset 1	
0 to 1	1	10.00%
1.1 to 5	4	50.00%
5.1 to 10	5	100.00%
10.1 to 50	0	100.00%
> 50	0	100.00%
	Dataset 2	
0 to 1	1	10.00%
1.1 to 5	4	50.00%
5.1 to 10	4	90.00%
10.1 to 50	1	100.00%
> 50	0	100.00%

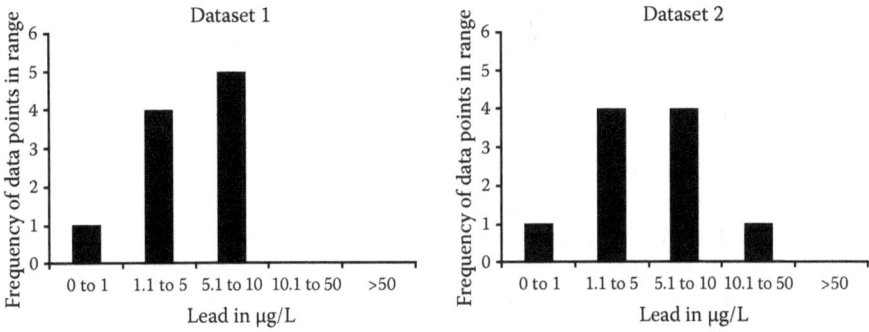

FIGURE 5.1   Histograms of two datasets in Table 5.1 using frequency of data points in each range.

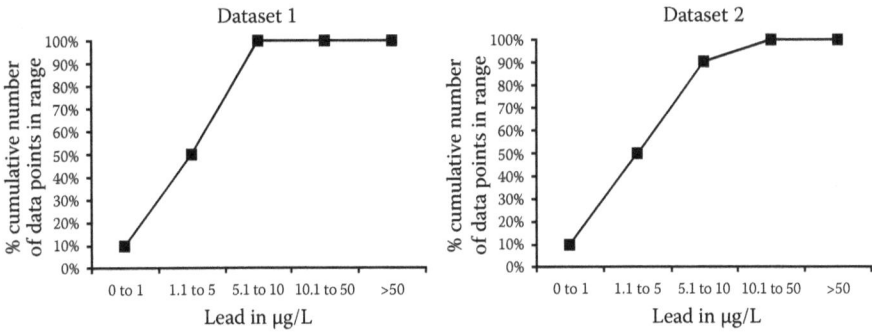

FIGURE 5.2   Histograms of two datasets in Table 5.1 using % cumulative number of data points in each range.

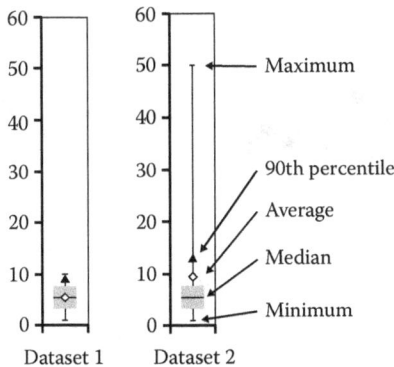

FIGURE 5.3   Box and whisker plots of two datasets in Table 5.1.

encompasses the two inner groups with the black dividing line denoting the median (50th percentile) of the dataset. The highest group includes values between the top of the grey box and the top "whisker"—the vertical black line with a short, horizontal line as the maximum data-point value. The lowest group includes values between the bottom of the grey box and the bottom "whisker." The diamond shape locates the average value of all the data points. The black triangle locates the 90th percentile of all the data points. Box and whisker plot graphing programs can be found on the internet as add-ins for common spreadsheet programs. Refer to Table 5.1 for the minimum, median, average, 90th percentile, and maximum values plotted in Figure 5.3.

The distribution of data can give important information. In the example shown in Table 5.1, a higher maximum value in a dataset increased the 90th percentile value. For the Lead and Copper Rule, this can be a matter of compliance versus noncompliance. The reason for the change in maximum value would need to be investigated.

5.4   TIME SERIES GRAPHS

The most common graph for distribution system water quality data is a graph of results plotted over time (Figure 5.4). This is called a "time series." The graph should tell us: Are current results similar to previously collected results or are they different? Look for the range within which the data values hover and get to know the typical range for that parameter. Is it a tight range or a wide range?

In Figure 5.4, data start high and fall to a lower range. Then, the data follow various patterns after that.

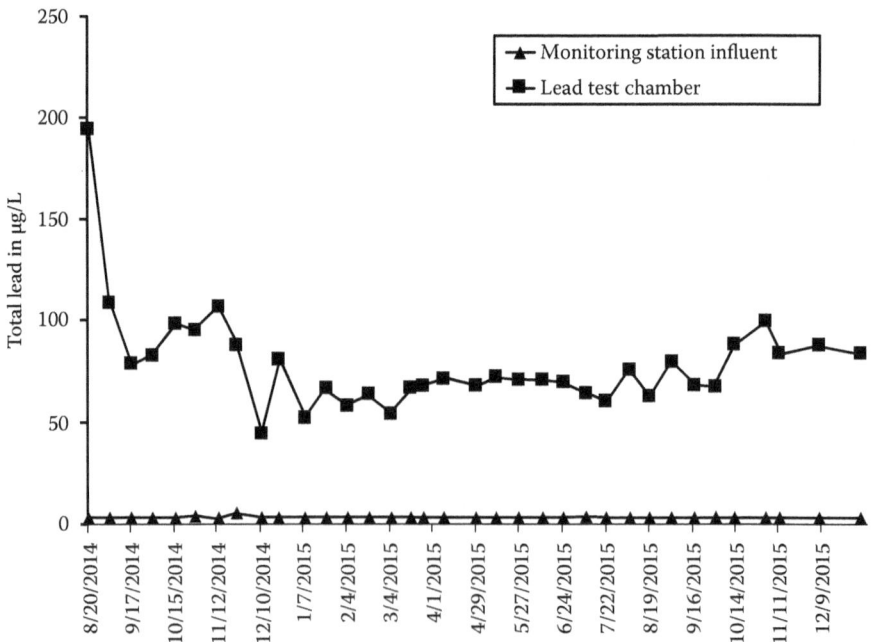

FIGURE 5.4   Example time-series graph showing total lead concentration.

5.5 SHEWHART CONTROL CHARTS

The time series graph is sufficient for a person to study the trends of the data. However, there is a better method by which to study trends of data over time. That method was developed in the 1920s by Dr. Walter Shewhart of Bell Laboratories (Wheeler and Chambers 1992).

A Shewhart Control Chart can assist in determining which data points on a time-series plot are atypically high and which are atypically low. It can also identify other water quality trends indicating a changing water system. It is a data analytical technique that has been borrowed for water quality and water system data (Cantor and Cantor 2009; Cantor et al. 2012) from the field of industrial process control and improvement, called statistical process control.

The control chart is essentially a graph of data over time. Ease of interpreting the data is achieved by the guidelines plotted on the graph, which are plots of the Shewhart statistics calculated from the dataset. The statistics describe the expected variation of the data if the system remains operating under its existing conditions. Data points that fall outside of the lines of expected variation show that new factors are influencing the system. Other data patterns on the chart can show that changes to the status quo are beginning to occur (Wheeler and Chambers 1992).

Figure 5.5 summarizes the control chart's features. Here, monitoring data are plotted over time. The average is drawn as a line through the data. A unit of variation, called a sigma unit, is calculated and then used to define the range (±3 sigma units around the average) in which 99% of the data are expected to fall.

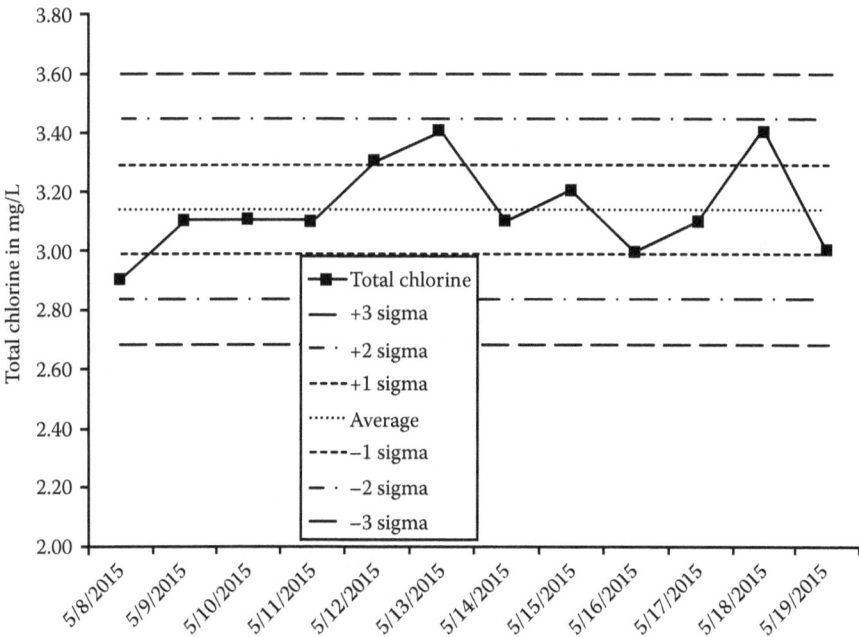

FIGURE 5.5 Summary of Shewhart Control Chart characteristics.

Standard deviation is a type of a sigma unit. However, standard deviation can only be used on data from randomized experiments where data points are independent of each other. In water quality monitoring, samples can only be taken sequentially over time like on an assembly line; the conditions at a previous sampling time might affect the conditions at the next sampling time. Therefore, the more general sigma unit is used. Refer to Wheeler and Chambers (1992) for a better understanding of the technique.

For Figure 5.5, the following are defined:

- Average + 3 sigma units = Upper Control Limit = UCL
- Average – 3 sigma units = Lower Control Limit = LCL
- Between the UCL and the LCL = the expected range of the data. This is the range that will result from routine factors working on the system and 99% of the data points are expected to fall.

Patterns of atypical data can be found on the graph when

- Data fall outside the three sigma unit lines.
- At least two out of three successive values fall on the same side of the average and are two sigma units or greater away from the average.
- At least four out of five successive values fall on the same side of the average and are one sigma unit or greater away from the average.
- Eight or more successive points fall on the same side of the average line.

Any data following the above patterns should trigger an investigation of system operations to determine the cause. If results are unfavorable, then corrections can be made to system operations. If results are favorable, then the continuation of such operating conditions could be beneficial.

Figure 5.6 shows the time series graph of Figure 5.4 as a Shewhart Control Chart. The first data point is above the upper control limit denoting a special circumstance that does not match the rest of the data. The first eight data points are above the average, while the next eight-plus data points are below the average. These two sets of data points delineate two very different conditions occurring in the distribution system or the test chamber.

Using Drs. Shewhart's and Deming's statistical process control techniques does not require an in-depth background in statistics. The construction and use of a control chart are not complicated. There are also software packages that can create the charts.

Be aware, though, that while these charts are easy to use, they are also easy to misuse. It is important to understand certain nuances about the technique. First, the way that the Shewhart statistics are calculated will depend on the frequency of sampling for a specific parameter at a specific site. If products on an industrial assembly line were measured for certain characteristics every 15 minutes, the Shewhart Control Chart technique could be used to group the four parameters taken each hour and use their average value and range instead of their individual values in the statistics calculation. A water utility using an online sensor for measurement of a

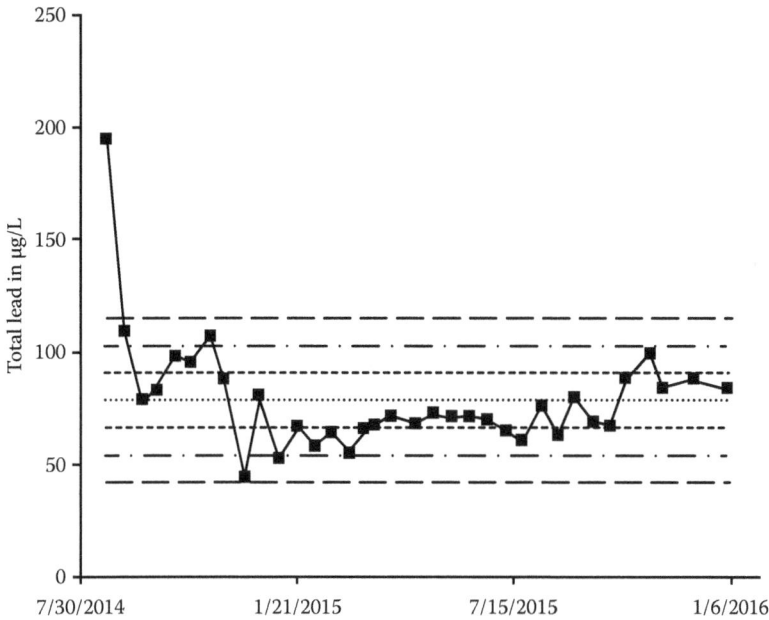

FIGURE 5.6 Total lead released in the lead test chamber from Figure 5.4 graphed as a Shewhart Control Chart.

water quality parameter may also choose to group data taken every 15 minutes by hour or possibly group all measurements by day. The decision as to how big a data group should be is up to the analyst. The decision is based on what questions are to be answered with the data and what degree of sensitivity is desired on the control chart. "Subgroups must match the time period which characterize the same changes in process" (Wheeler and Chambers 1992). For people who want to answer important questions for frequently sampled measurements, more in-depth reading regarding the Shewhart Control Charts in statistical process control references can help (Wheeler and Chambers 1992).

There are also situations—especially in water systems—where frequent measurements cannot be made. Data from a pipe loop-style apparatus in the distribution system as described in Chapter 4 are obtained infrequently—once a week, every other week, once a month. Data from some regulatory site sampling, such as for the Total Coliform Rule as described in Section 3.4, are obtained infrequently at each regulatory site, from a few times a month to every few months. Therefore, so much time passes between measurements along with changes that may be occurring in the distribution system that it does not make sense to group one data point with others; they all represent a different state of the distribution system. In this situation, data are not to be grouped. "Each value is uniquely identified with a specific period of time. This is a logical subgroup of size of n = 1" (Wheeler and Chambers 1992). Therefore, an "individuals" chart should be calculated.

Another nuance concerning Shewhart Control Charts is whether to use the average of data versus the median of data. Some argue that using the average of data creates a control chart with possibly very large variations of data that may not be representative of typical operations and, therefore, median values should be used to smooth out the graphs (Cornwell et al. 2015). Others warn that "the chart for the subgroup medians will never be as sensitive as the regular control chart for subgroup averages. It will always be slower in responding to changes in the process. [The subgroup averages chart] should always be preferred over a median chart" (Wheeler and Chambers 1992). A high variation in a dataset is a signal to identify the influencing factors in a water system, whether it is a faulty sampling or analytical method or an actual system operations issue. Dismissing high variation as "noise" or "outliers" is a lost opportunity to understand and improve a process.

In the monitoring strategies described in this chapter, the goal is to respond to water quality trends and determine what is pushing the water system to a different state. It is true that sometimes the trends on a sensitive, average-based Shewhart Control Chart might be false alarms. However, by responding to each alarm, even seemingly minor issues, such as sampling protocols, problems with a chemical feed pump, or problems with operator protocols, can be corrected and those factors removed from keeping a parameter within statistical control in a distribution system.

Table 5.3 depicts an example of how to calculate an individual's average-based chart (subgroup size n = 1). If a software package is not available to calculate the Shewhart Control Chart statistics and graph them, this demonstration can be set up in a spreadsheet in order to work with other data.

TABLE 5.3
Calculation of an Average-Based Shewhart
Control Chart with Subgroup Size n = 1

Date	Total Chlorine in mg/L	Moving Range
5/8/2015	2.90	
5/9/2015	3.10	0.20
5/10/2015	3.10	0.00
5/11/2015	3.10	0.00
5/12/2015	3.30	0.20
5/13/2015	3.40	0.10
5/14/2015	3.10	0.30
5/15/2015	3.20	0.10
5/16/2015	3.00	0.20
5/17/2015	3.10	0.10
5/18/2015	3.40	0.30
5/19/2015	3.00	0.40
Average	**3.14**	**0.17**

In Table 5.3, data are listed for total chlorine at one sampling site in the order that they were measured. The average of the data is calculated as 3.14. Then, the "moving range" is calculated by subtracting the value of one result from the previous result. The absolute value of the difference is used so that there are no negative numbers. Finally, the average of the moving range is calculated as 0.17. The moving range provides an understanding of the variation from one measurement to the next.

The moving range is used to calculate the variation of the data in the dataset. A unit of variation is called a "sigma unit," as explained in Figure 5.5. The sigma unit is calculated by dividing the average moving range by a special constant based on data subgroup size. The constants have been calculated based on statistical concepts, but like many other concepts in statistics, the constants are conveniently listed in tables for use in constructing control charts (Wheeler and Chambers 1992). For the case of a dataset with a subgroup of size n = 1, the constant is 1.128. In the example of Table 5.3, the sigma unit of the dataset is 0.17 divided by 1.128, which equals 0.15.

Now, there is enough information to construct the "guidelines" (calculate the Shewhart Control Chart statistics) behind the time-series graph of total chlorine measurements. Figure 5.7 shows the Shewhart Control Chart for the data in Table 5.3. The calculations for the guidelines are as follows:

- Average = 3.14
- Sigma unit = 0.17/1.128 = 0.15

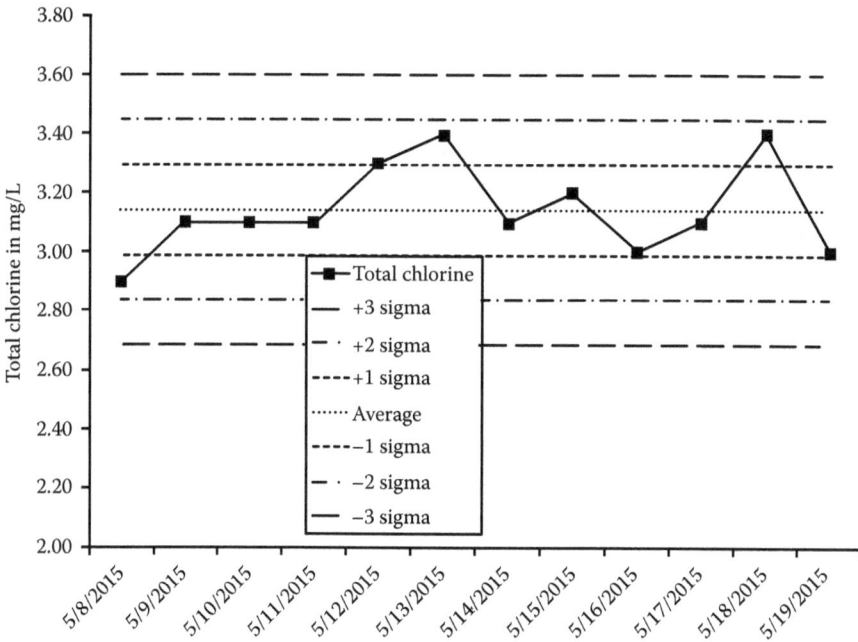

FIGURE 5.7   Shewhart Control Chart of Table 5.3 data.

- -1 sigma unit $= 3.14 - (1 \times 0.15) = 2.99$
- -2 sigma unit $= 3.14 - (2 \times 0.15) = 2.84$
- -3 sigma unit $= 3.14 - (3 \times 0.15) = 2.68 =$ Lower Control Limit
- $+1$ sigma unit $= 3.14 + (1 \times 0.15) = 3.29$
- $+2$ sigma unit $= 3.14 + (2 \times 0.15) = 3.45$
- $+3$ sigma unit $= 3.14 + (3 \times 0.15) = 3.60 =$ Upper Control Limit

5.6   SHEWHART STATISTICS COMPARISONS

The Shewhart Control Chart statistics can also be used to compare results between sites and between water systems. The average of the dataset, the Upper Control Limit (UCL), and the Lower Control Limit (LCL) are used for these comparisons. As stated previously, the UCL equals the average value + 3 sigma units; the LCL equals the average value − 3 sigma units. The difference between the UCL and the LCL is the range where 99% of the data points are expected to fall. A table can be constructed for each water quality parameter where the UCL, average, and LCL for each sampling site are listed, as shown in Table 5.4. The UCL and the LCL should not be mistaken for the actual maximum and minimum results observed. Instead, these are statistical values, as described previously.

Taking this comparison between sites a step further, these values were graphed as in Figure 5.8.

A method for comparing disinfection data and turbidity data between Total Coliform Rule sites was described in Section 3.4. In those examples, minimum and maximum values were used to compare variation from site to site. Instead of minimum and maximum values, the Lower Control Limit and Upper Control Limit can be used. But make sure the nature of the "whiskers" is identified on the graph. The minimum and maximum values have actually been measured at each site. The UCL and LCL values bound a range in which measured values have a 99% chance of falling. These values may not have actually been measured at the site.

TABLE 5.4
Shewhart Control Chart Statistics Summary of Two Sampling Sites

Total Lead: Comparison of Shewhart Control Chart Statistics by Site in µg/L

Site	UCL	Average	LCL
Monitoring Station Influent	3.4	3.1	2.7
Lead Test Chamber	115.0	78.5	42.1

Note: This is a table of the statistics representing average value and Shewhart Control Chart statistical variation. This does not represent actual maximum and minimum results observed.

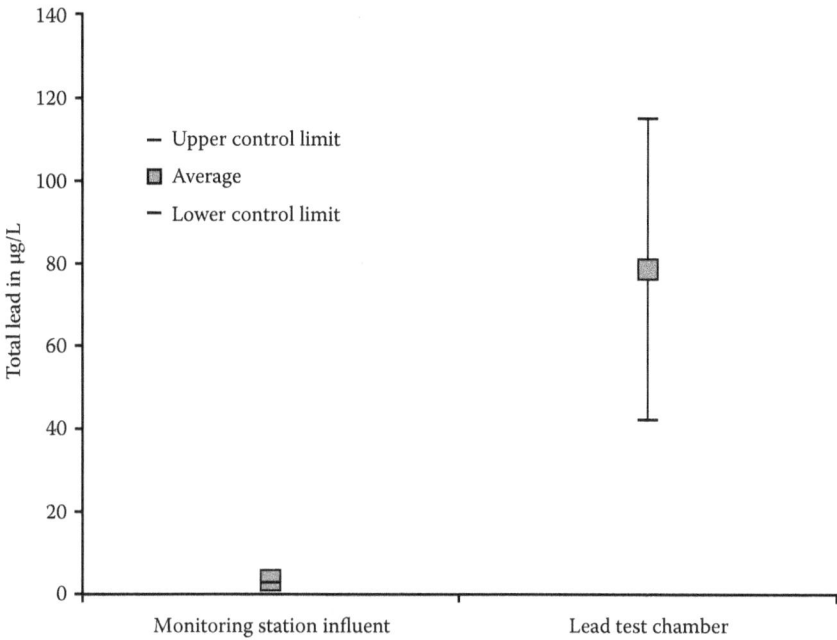

FIGURE 5.8   Graph of Table 5.4 Shewhart Control Chart statistics summary.

5.7   CORRELATIONS

So far, the discussion on analyzing data has focused on studying one water quality parameter and determining how it relates to the same parameter measured at a different time or a different location. In this section, correlation techniques are discussed to determine the relationship between two different water quality parameters.

5.7.1   REGRESSION TREND LINES

A common means of determining the correlation between two parameters is to plot them on a graph where the x axis is Parameter 1 and the y axis is Parameter 2, measured at the same time from a specific site. Then, the x and y data pairs are used to calculate factors in a mathematical equation that relates x to y. For example, a linear regression correlation uses a function for a line, where y = ax + b. The factors a and b are calculated using the actual x and y pairs of the measured data. A correlation coefficient is also calculated to determine how well the line equation estimates the actual x and y relationship. If the data tend to form this linear relationship, the correlation coefficient, "r^2," will approach 1.0. If there is no tendency to form a linear relationship, r^2 will approach 0.0. Figure 5.9 displays two parameters that estimate a line with a correlation coefficient of 0.61. The two parameters have a high degree of correlation but not a perfect correlation.

Other functions can be used to estimate a relationship between parameters. Figure 5.10 shows two water quality parameters correlated by an exponential

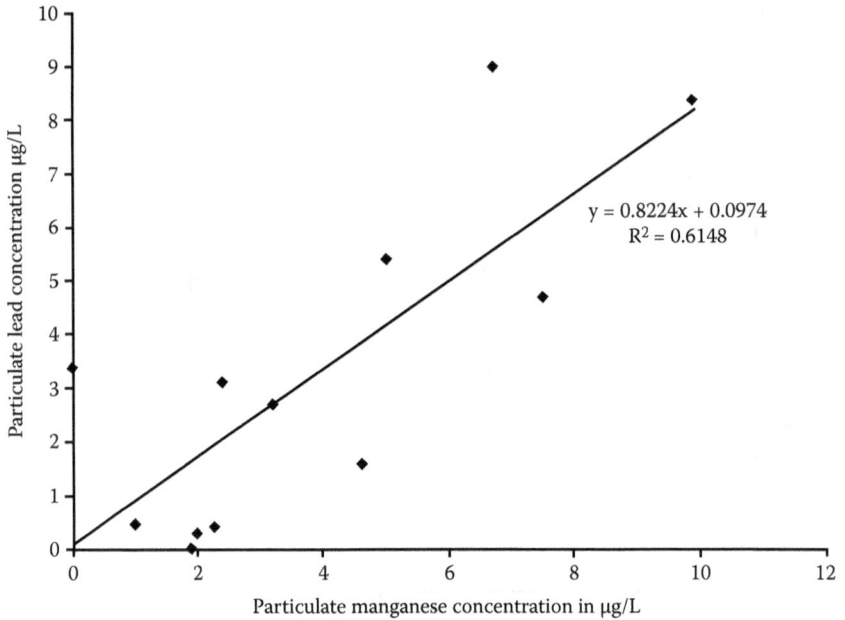

FIGURE 5.9 Example of linear regression correlation between two water quality parameters.

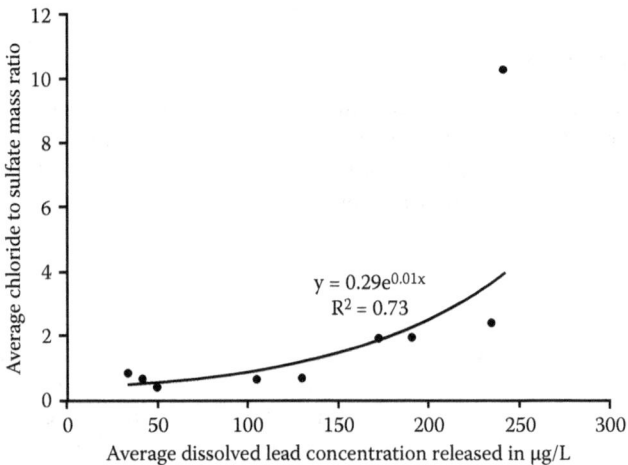

FIGURE 5.10 Example of exponential regression correlation between two water quality parameters.

function where $y = ae^{bx}$, where a and b are constants calculated using the x and y data pairs.

A negative correlation coefficient means that the two parameters are trending inversely to each other. That is, when one parameter is increasing, the other is decreasing.

The calculation and plotting of regression lines is a function on Microsoft Office Excel spreadsheet graphs.

5.7.2 SPEARMAN RANK CORRELATIONS

Water quality parameters do not typically correlate with each other so neatly into set equations. Correlation of such unruly data can be determined using the Spearman Rank Correlation technique. This is appropriate for water quality monitoring data because the technique can be applied to "nonparametric" data where one data point could influence a subsequent data point. The data from water systems are not considered independent and random and, therefore, could influence a subsequent data point.

The Spearman correlation is a technique that determines if two parameters are always increasing at the same time. If they are always increasing at the same time, they would have a perfect Spearman correlation coefficient of 1.0. If they increase together most of the time, they have a coefficient of a fraction of 1 and can range down to 0. The correlation also can discern if one parameter always increases while the other always decreases. If they always do this, they have a coefficient of −1. Fewer occurrences together would give negative fractions between −1 and 0.

Conclusions drawn from the Spearman correlation coefficients or any correlation technique should be made carefully. Common trends between two parameters do not prove causation. The two parameters may, instead, be characteristics of some other phenomena. For example, if lead release is trending with alkalinity of water, one should not assume that the lead release is the result of the increasing alkalinity. Instead, for example, an operational change of source water may be the real factor that increases both the alkalinity and the lead release. There can be many possible explanations that require more system study to decipher.

In Microsoft Office Excel, a Pearson correlation coefficient for parametric data can be calculated and extra steps can be taken to transform this calculation into a Spearman rank correlation coefficient for nonparametric data. A description can be found on internet discussions. Other statistical software packages can perform the Spearman's rank correlation directly.

5.7.3 SPARKLINES

Time-aligned graphs of water quality parameters should also be inspected along with the correlation results. Sparklines are useful in this case. They are graphs without x and y axis labels that show the shape of the line connecting the data and the general trends of the data. When sparklines are created with data of multiple parameters over the same time period, they can be compared against each other to determine if trends are similar. Figure 5.11 is an example of sparklines that display two water quality parameters trending inversely with each other.

There are other reasons to inspect sparklines and other graphs instead of fully depending on Spearman Rank or other correlation coefficients. A trend pattern of one water quality parameter may be repeated in another water quality parameter at a later time period. Or, two water quality parameters may trend together only during

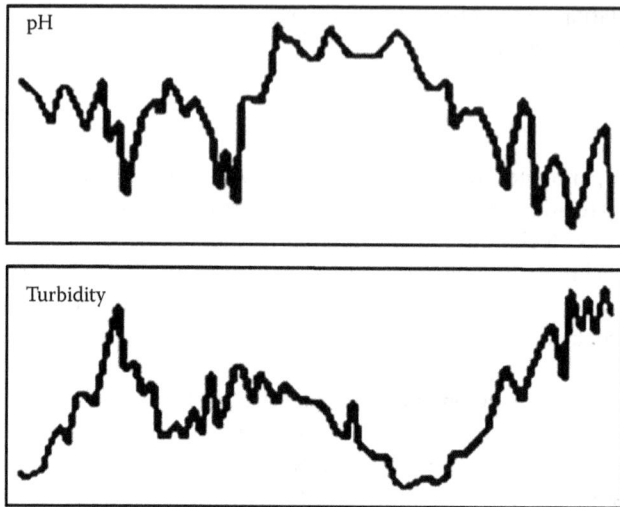

FIGURE 5.11 Example of sparklines for data trend comparison.

a specific time period and trend oppositely or not at all in a different time period. These subtleties would be missed with a calculated correlation coefficient.

Microsoft Office Excel can create sparklines.

5.8 AREA CHARTS

There are some water quality parameters that can be broken into fractions. For example, total lead is composed of dissolved lead and particulate lead; total iron is composed of dissolved iron and particulate iron; total chlorine is composed of free chlorine and combined chlorine. It is informative to compare the fractions visually. This can be accomplished with a bar chart or a shaded-area graph.

Shaded-area graphs are similar to bar charts where the two fractions add up to the total concentration. Instead of bars, data points for one fraction are connected by a smooth curve over time and those for the other fraction are connected to each other as well. The area between the two curves is shaded, giving a dramatic visual effect for comparing the quantities of the two fractions. See Figure 5.12. These graphs can be created in Microsoft Office Excel.

5.9 SUMMARY

Several data analysis techniques have been demonstrated in this chapter. In summary, when studying one water quality parameter at one sampling site, the following techniques are informative:

- Comparison of distributions of data between time periods of interest, such as quarterly or annually. See Figures 5.1 through 5.3.

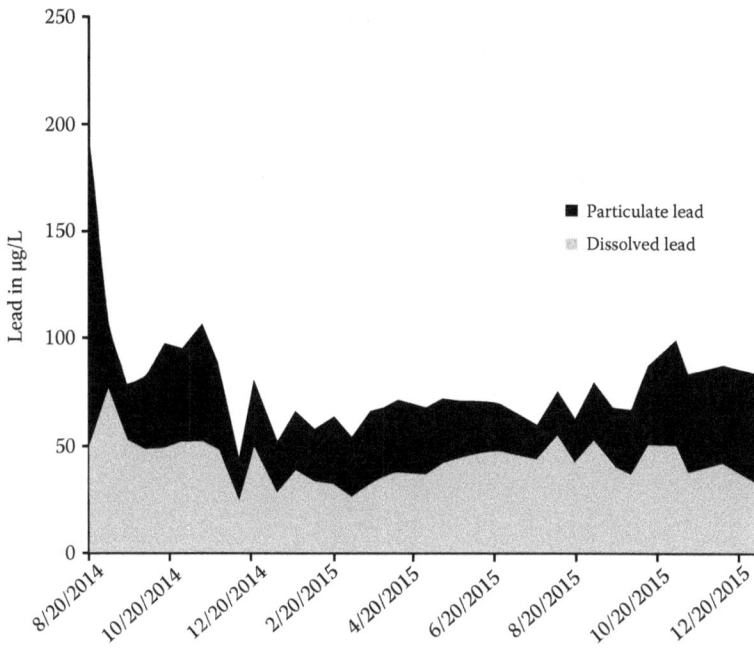

FIGURE 5.12 Example of shaded-area graph showing dissolved and particulate lead concentrations at a sampling site.

- Calculation of Shewhart Control Chart statistics and plotting on a time-series graph. Interpretation of the trends based on Shewhart Control Chart rules and taking action if trends are leading to bad outcomes. See Figures 5.5 and 5.7.

When studying one water quality parameter at multiple sites:

- Plotting of data per site on the same time-series graph for visual comparison. See Figures (3.6, 3.7, and 5.4).
- Comparison of distributions of data by site. See Figures 5.1 through 5.3.
- Calculation and comparison of Shewhart Control Chart statistics in a table or in a comparative graph. See Figure 5.8.

For multiple water quality parameters at one site:

- Running Spearman rank correlations between all water quality parameter pairs. See Section 5.7.2.
- Plotting and comparison of sparklines for each water quality parameter. See Figure 5.11.

- If two parameters appear to follow each other in a consistent orderly fashion: plotting of one parameter against the other and calculation and plotting of regression lines. See Figures 5.9 and 5.10.

One other data analysis graphing technique was described for data that represent fractions of a total concentration, such as where dissolved lead and particulate lead concentrations equal the total lead concentration or where free chlorine and combined chlorine equal the total chlorine concentration. In these cases, shaded-area graphs are informative. See Figure 5.12.

6 Case Studies

This book is filled with protocols and other information about how to carry out a water distribution system monitoring program and make improvements in the water system. In this chapter, three case studies demonstrate these methods. Water Systems 1 and 2 were studied during the Water Research Foundation Project 4586, and their data and graphs are derived from that report (Cantor 2017).

6.1 WATER SYSTEM 1

Water System 1 was originally a groundwater system from 1886 to 1957. By 1957, a transmission line had been built to bring lake water to a central treatment facility and the water system switched to the surface water source.

When the water system exceeded the Lead and Copper Rule Lead Action Level of 15 µg/L in 2011, an investigation was performed that used existing information in order to hypothesize the possible causes of the lead increase.

The existing information included the configuration of the distribution system, especially regarding lead and galvanized iron water service lines (Section 3.1). Water System 1 information, in Table 3.1, shows the prevalence of lead water service lines connected to privately owned galvanized iron water service lines. This was most likely a significant factor related to elevated lead at consumers' taps. The adsorption and accumulation of lead on galvanized iron pipe scales with eventual crumbling of the lead-laden scale into the water has been acknowledged in the technical literature as a major form of lead transport in premise plumbing (McFadden et al. 2011). Plans were made for lead and galvanized iron service line removal.

The existing information also included the harvesting of a lead service line for chemical scale analysis (Appendix B). An analysis of the lead service line scales found crumbly scales of iron, manganese, and lead. Iron and manganese scales on pipe walls have been found to adsorb, accumulate, and transport lead to consumers in water systems (Schock et al. 2014). Only standard hydrant flushing had been performed over the years to clean water mains; the pipe wall chemical analyses showed that this cleaning activity had not been sufficient to remove system debris most likely deposited decades ago when groundwater high in iron and manganese was used. An engineered unidirectional (high velocity) flushing program was designed and plans were made to carry it out for this water system.

Existing information also included a timeline of system changes, as discussed in Section 3.2. See Table 3.2 for the Water System 1 timeline. In addition, studying Water System 1's Lead and Copper Rule data was informative. Figure 6.1 shows the Lead and Copper Rule data through 2012. Refer to Section 3.3 for construction of the graphs. For lead, the 90th percentile lead was measured greatly higher than the Action Level in 2011. Before the large lead concentration change, the lead levels had slowly been increasing since around 2000. For copper, the 90th percentile concentration did

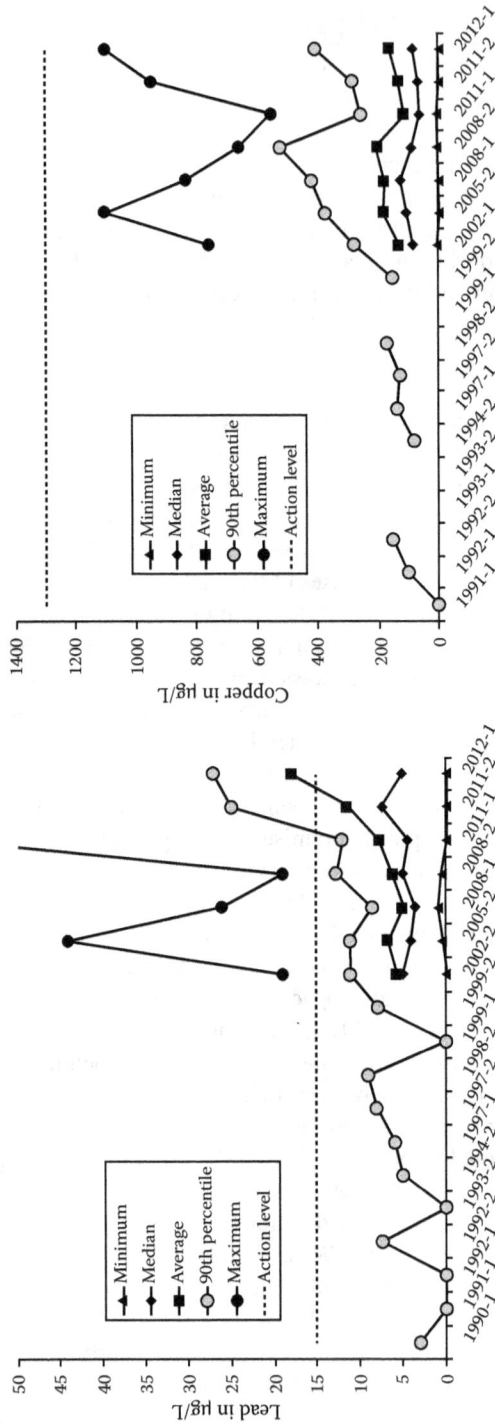

FIGURE 6.1 Lead and Copper Rule data through 2012 for Water System 1.

not exceed the Action Level, but copper levels had also been increasing since around 2000. According to the timeline, the ozone disinfection was installed in 2000. The water utility personnel had stated that there was an increase in corroded valves and meters in the distribution system just after the ozone process was placed online. A new large transmission line was installed in 2005 and repairs were made to water system components and the older transmission line in 2010 and 2011. These were the major events that led up to the Lead and Copper Rule compliance issue in 2011. Given the possible increase in microbiologically accessible organic carbon from the ozone process (Escobar and Randall 2001) and the disturbance of existing biofilms from the transmission line installation and repairs, along with a reported increase in corroded valves and meters, a microbiological component to lead and copper release was suspected. Plans were made for a separate study of water system biostability.

Finally, plans were made to set up and carry out a comprehensive distribution system monitoring program as the water mains were cleaned of iron, manganese, and biofilms.

6.1.1 TOTAL COLIFORM RULE SITE DATA

The Total Coliform Rule site disinfection data and the additional turbidity data were not studied during the initial investigation for Water System 1, but that would have

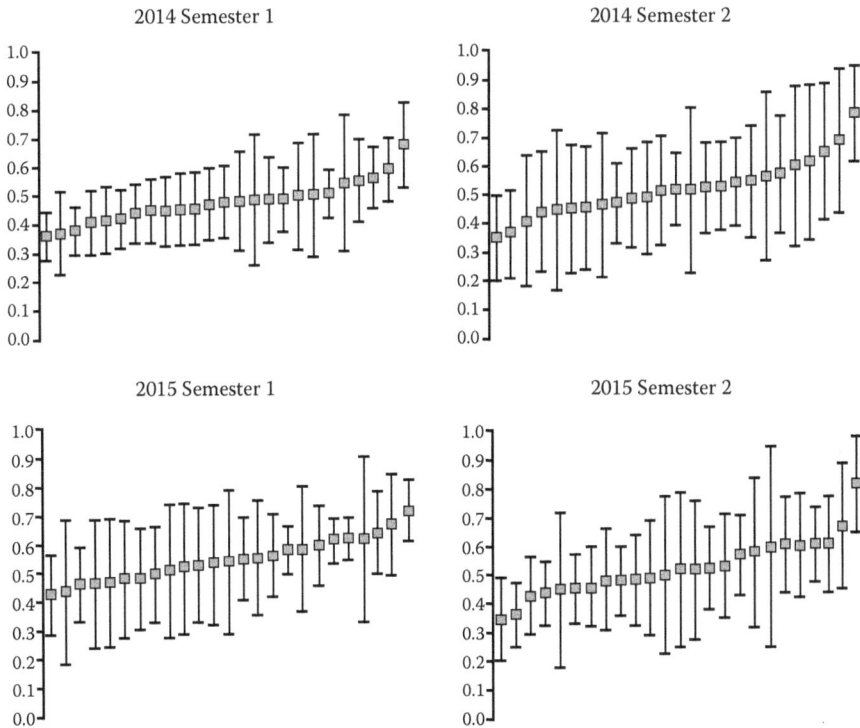

FIGURE 6.2 Water System 1: comparison of total chlorine in mg/L at Total Coliform Rule sites.

been helpful in pinpointing specific locations in the distribution system to investigate further. Use of Total Coliform Rule data is discussed in Section 3.4.

The data were plotted, however, during the comprehensive monitoring period and showed certain patterns, as seen in Figures 6.2 through 6.5.

Figure 6.2 shows that chlorine concentrations were slightly higher in 2015 than during the first half of 2014. In the second half of the year, the warmer half, disinfection concentration underwent more variability than in the colder half of the year. When the x-axis site labels are included on these graphs, specific locations can be investigated for lower disinfection levels and higher variability. Such an investigation can lead to insight into why water quality becomes degraded in that locality.

Figure 6.3 illustrates the turbidity (release of particulates into the water) in the distribution system network. As with disinfection concentration, specific sites can be pinpointed for further study. Variability in turbidity increased in the second half of 2014 and in 2015, with more variability occurring in 2015. The high velocity water main flushing occurred in the summers of 2014 and 2015 and the particulate variability could be related to this effort in cleaning the system, an ironic reality in cleaning up water systems.

Figure 6.4 shows the overall range of disinfection concentrations at all sites and their patterns over time. These graphs can identify general trends around the distribution system network and can locate the degree to which some sites might not be

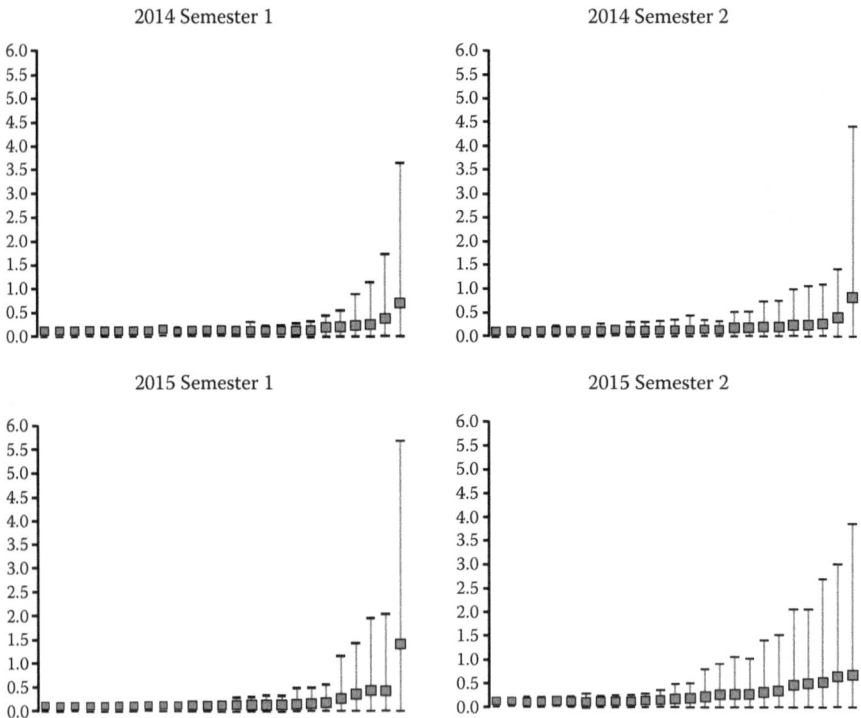

FIGURE 6.3   Water System 1: comparison of turbidity in NTU at Total Coliform Rule sites.

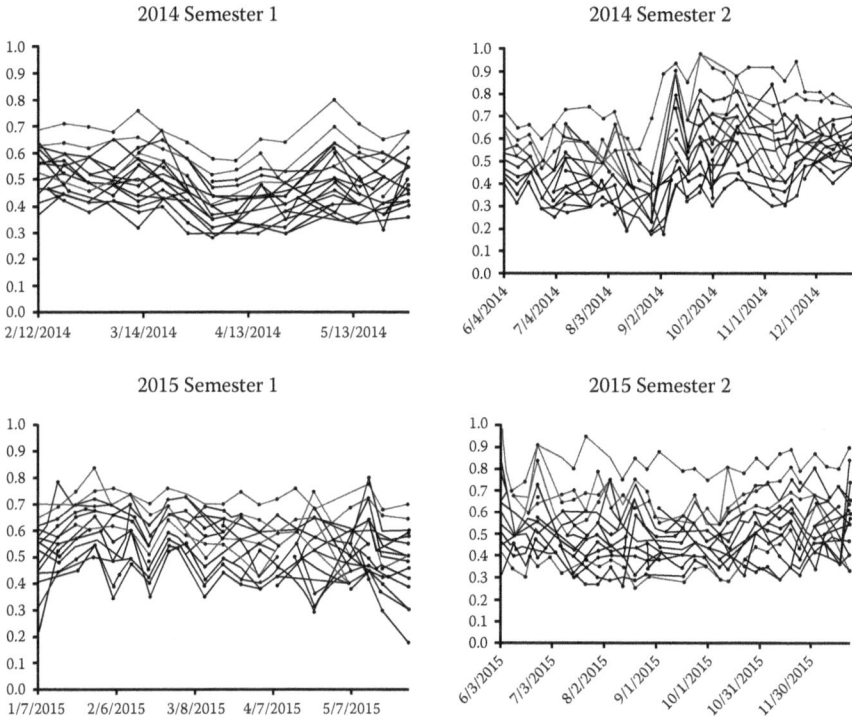

FIGURE 6.4 Water System 1: comparison of total chlorine in mg/L at Total Coliform Rule sites over time.

following the general trends. It can be seen that disinfection concentration increased, in general, from September 2014 through the end of the year. There were greater differences in disinfection concentration around the distribution system in summer, in general.

Figure 6.5 shows that turbidity changes were typically large singular events and not smoothly changing like the disinfection concentration. Turbidity in the second half of the year was greatly higher than in the first half. As mentioned previously, the reason could be related to high velocity water main flushing efforts as well as interactions that occur in warmer temperatures.

6.1.2 RESIDENTIAL SAMPLING

During the comprehensive monitoring program, two residences were sampled multiple times using the profile sampling method described in Section 4.7.

The characteristics of the water flowing into each residence, B1 and B2, are shown in Table 6.1.

Total lead concentrations as they change from the first liter at the kitchen tap to the tenth liter in the water main are shown for Site B1 in Figure 6.6. More information can be seen in Figure 6.7, where the total lead has been differentiated into

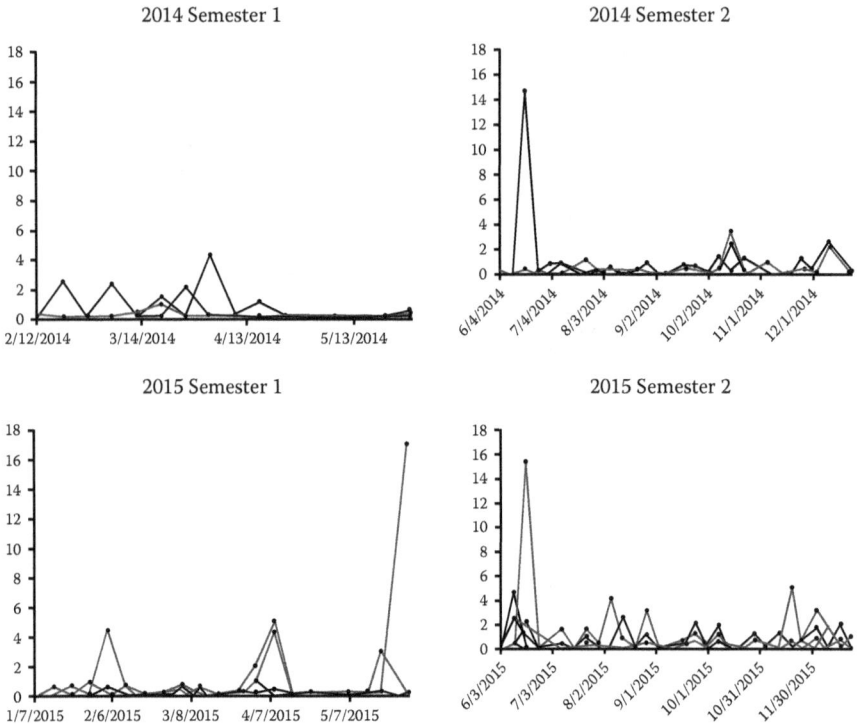

FIGURE 6.5 Water System 1: comparison of turbidity in NTU at Total Coliform Rule sites over time.

dissolved and particulate lead. It can be seen that the lead was mostly in dissolved form. In addition, the water service line was flushed at high velocity after the first sampling. Instead of lowering the lead levels, they increased and never subsided to the original levels. This was surprising in that, in other water systems, higher lead levels are often because of release of particulate lead from pipe walls and high velocity flushing to a lower turbidity is often helpful in lowering lead concentrations. This is an example as to how dangerous it is to make assumptions and not confirm them; every water system is different.

For Site B2, the first sampling was performed before unidirectional flushing of water mains occurred nearby on 9/24/2014. Subsequent sampling showed elevated dissolved lead, but all levels had fallen to lower concentrations by the final sampling on 12/7/2015. The elevated lead concentrations never exceeded 15 μg/L. See Figures 6.8 and 6.9.

Copper concentrations in profile sampling are typically highest near the kitchen sink sampling tap because residences commonly have more copper at that location. This was true for Sites B1 and B2. Neither of the residences sampled exhibited extreme copper concentrations.

Iron, manganese, and aluminum were not detected to a significant level in either of the residences.

TABLE 6.1
Water System 1: Residential Sampling Flowing Water Characteristics

	B1	B2
Aug/Sep 2015		
Calcium in mg/L	34.2	33.6
Magnesium in mg/L	11.4	11.4
Chloride in mg/L	13.6	13.6
Fluoride in mg/L	0.63	0.61
Phosphorus in mg/L as P	*ND	ND
Orthophosphate in mg/L as P		
Potassium in mg/L	1.3	1.28
Silica in mg/L as SiO_2	2.19	2.15
Sodium in mg/L	8.15	8.12
Sulfate in mg/L	21.8	21.9
Total Alkalinity in mg/L as $CaCO_3$	110	110
Nov/Dec 2015		
Calcium in mg/L		
Magnesium in mg/L		
Chloride in mg/L	15.5	15
Fluoride in mg/L	0.73	0.71
Phosphorus in mg/L as P		
Potassium in mg/L		
Silica in mg/L as SiO_2		
Sodium in mg/L		
Sulfate in mg/L	23.5	23.7
Total Alkalinity in mg/L as $CaCO_3$	110	110

* ND = not detected; concentration below the laboratory limit of detection

The residential sampling called attention to a high degree of dissolved lead in the Water System 1 when iron, manganese, and aluminum particulates were not present.

6.1.3 PROGRAM TO MONITOR THE EXTREMES OF THE WATER SYSTEM

Section 4.1 encouraged a combination of strategies of site selection in order to obtain data from various perspectives for comparison. It recommended sampling a network of sites in the distribution system, which is described in Section 6.1.1 with the study of disinfection concentration and turbidity at Total Coliform Rule sites. Sampling at critical buildings in the system was recommended and was described for Water System 1 in Section 6.1.2. The final sampling strategy—sampling at extreme conditions in the water system—was also performed within a comprehensive distribution system sampling plan in Water System 1.

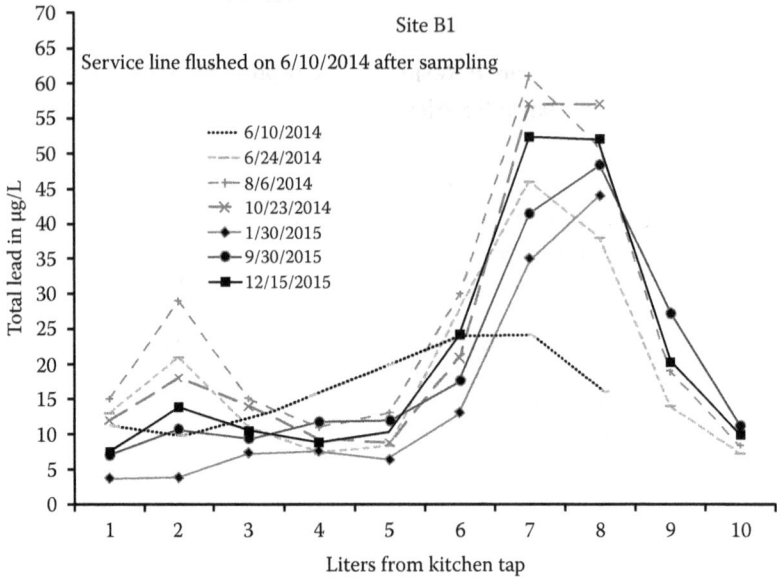

Liters from kitchen tap	Piping material associated with sample
1	Galvanized iron pipe
2	Galvanized iron pipe
3	26% of length galvanized iron and 74% copper pipe
4	Copper pipe
5	Copper pipe
6	4% of length copper and 96% lead pipe
7	Lead pipe
8	21% of length lead service line and 79% water main
9	Water main
10	Water main

FIGURE 6.6 Water System 1: residential profile sampling for Site B1 total lead concentrations.

One PRS Monitoring Station with one lead test chamber and one copper test chamber was located at a high water age location. The location where the treated water entered the distribution system water mains and tanks was also a flowing water sampling site. Table 6.2 summarizes the sampling sites.

Table 6.3 summarizes the monitoring plan, its sampling sites, water quality parameters, type of water sample (flowing or stagnant), frequency of sampling, and location of analysis (field or laboratory).

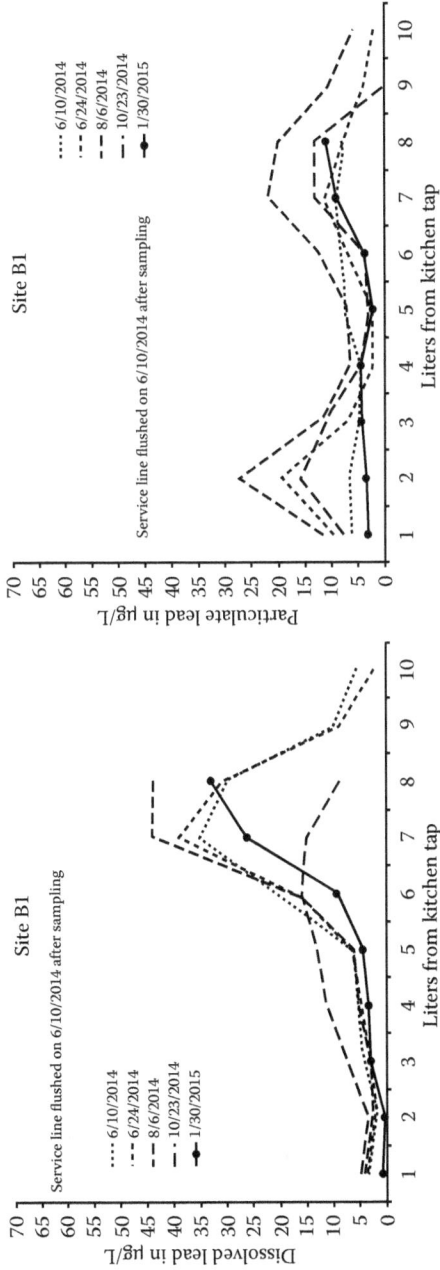

FIGURE 6.7 Water System 1: residential profile sampling for Site B1 dissolved and particulate lead concentrations.

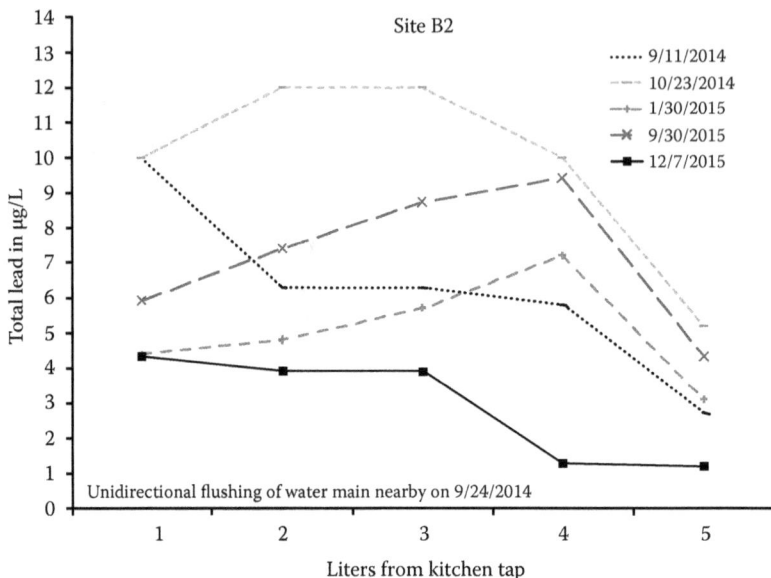

FIGURE 6.8 Water System 1: residential profile sampling for Site B2 total lead concentrations.

6.1.4 PRS MONITORING STATION RESULTS

Results of monitoring data pertaining to the data analysis techniques described in Chapter 5 are shown here. There are many more graphs and data analyses not shown in this example, but they were used on Water System 1 data. It is important to study the data from many different perspectives.

In general, the total lead in the flowing system water at the high water age location of the PRS Monitoring Station had an average concentration of 5.3 µg/L. The upper control limit (UCL) of the data variation (the highest expected concentration of lead 99% of the time) was calculated to be 15 µg/L. For copper, the average system water concentration was 11 µg/L, with the UCL at 33 µg/L. See Table 6.4.

The lead and copper concentrations released over time into the test chambers' stagnating water of the PRS Monitoring Station are shown in Figure 6.10. In these graphs, the black area represents the particulate fraction of each metal; the gray area represents the dissolved fraction. It is seen that the lead is mostly in dissolved form, as was found in the residential sampling described previously. There were also peaks

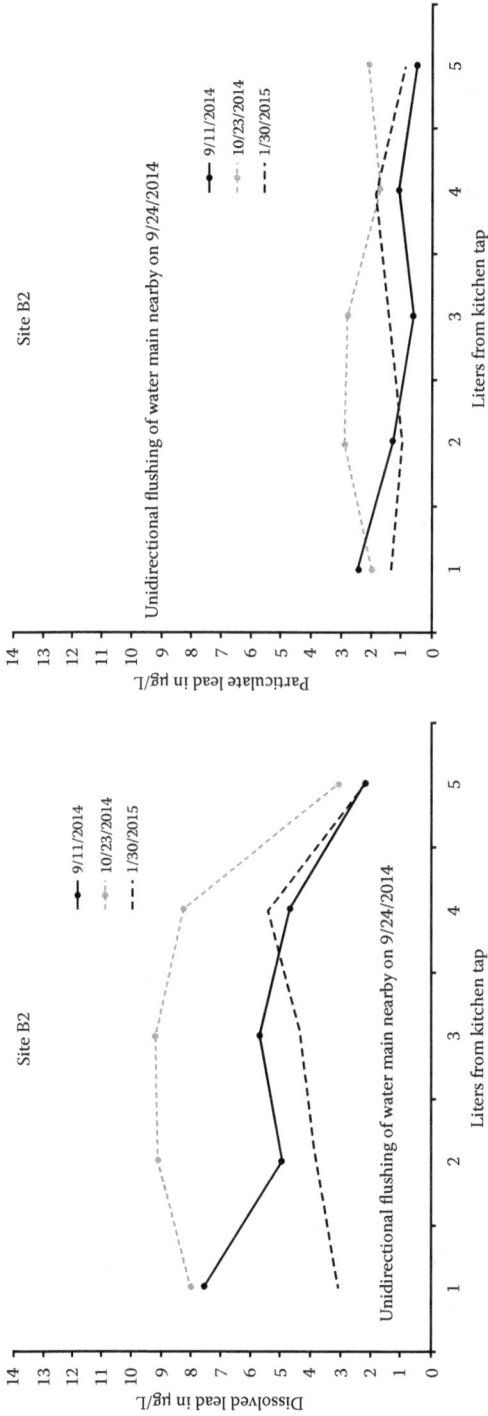

FIGURE 6.9 Water System 1: residential profile sampling for Site B2 dissolved and particulate lead concentrations.

TABLE 6.2
Water System 1: Sampling Sites Regarding a Comprehensive Distribution System Sampling Plan

Abbreviation	Description
EP	Flowing water at the entry point to the distribution system where treated water is pumped into the water mains and tanks
MS Inf	Flowing water at the influent sample tap of the PRS Monitoring Station, which was located at a high water age location; this site represents flowing system water characteristics at a high water age location
MS Pb	Stagnating water in the lead test chamber
MS Cu	Stagnating water in the copper test chamber

TABLE 6.3
Water System 1: Monitoring Plan

Frequency	Flowing Water Field Tests (Entry Point Water to the Distribution System and the Influent Water to the PRS Monitoring Station)	Flowing Water Lab Tests (Influent Water to the PRS Monitoring Station But Could Also Be Run on Distribution System Entry-Point Water)	Test Chambers' Stagnating Water Lab Tests
Weekly	Flow meter totalizer readings, pH, temperature, ORP, turbidity, conductivity, total chlorine, free chlorine		
Biweekly		Total metals scan, ATP	Total metals scan, dissolved metals scan, ATP
Monthly		Total alkalinity, chloride, sulfate, DOC	
Bimonthly		Total Phosphorus, Ammonia, Nitrite/nitrate	

TABLE 6.4
Water System 1: Total Lead and Copper Concentrations in Flowing System Water in µg/L Taken at a High Water Age Location (PRS Monitoring Station Influent Tap)

Metal	Highest Expected Concentration (UCL)	Average Concentration	Lowest Expected Concentration (LCL)
Lead	15	5.3	0
Copper	33	11	0

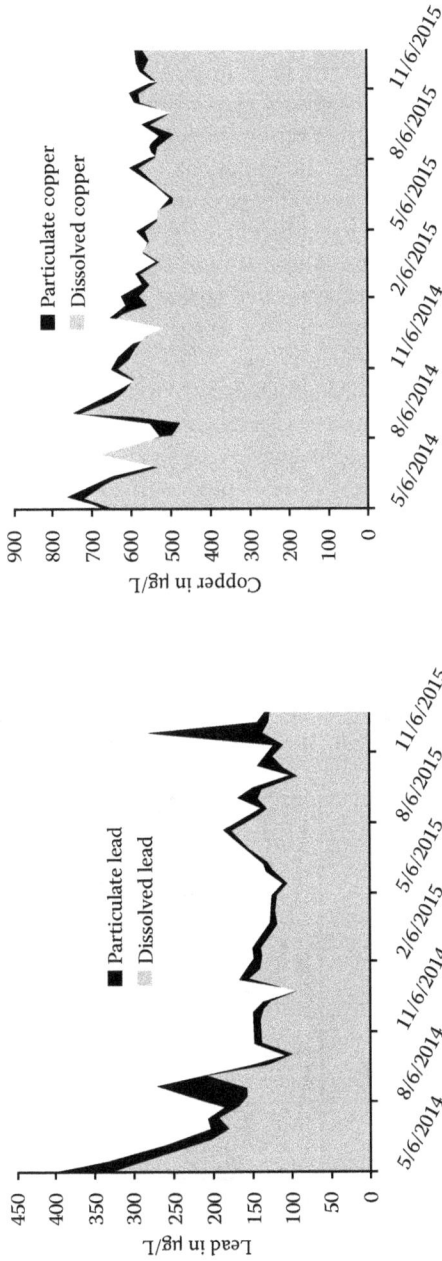

FIGURE 6.10 Water System 1: dissolved and particulate lead release into stagnating test chamber water in the PRS Monitoring Station.

of particulate lead at times. This is reminiscent of the random peaks of turbidity seen in the turbidity data of the Total Coliform Rule sites shown in Section 6.1.1. This was also predicted from the pipe wall scale analyses of a harvested lead service line.

Copper was also mostly in dissolved form, as shown in Figure 6.10. Copper did not experience the particulate release events that lead did.

As described in the comprehensive monitoring plan (Table 6.3), many water quality parameters were measured. Trends for each water quality parameter were studied to determine if any parameter's patterns appeared related to lead or copper release. This was an exercise in searching for correlations, as described in Section 5.7.

Special statistical software was used to perform a Spearman rank correlation test on all water quality parameters very quickly at once. In Chapter 5, it was noted that one also needs to inspect water quality parameter time-series graphs for a view of comparative trends in order to capture time-lagged common trends and trends that only occur at certain times of the year. The Spearman rank correlation calculations do not necessarily capture those conditions.

Chapter 5 also warned of conclusions drawn from successfully correlating the trends of two water quality parameters. "Correlation does not imply causation" is a common principle in the field of statistics. A person can only theorize about cause and effect between two parameters and continue testing the theory with more monitoring data and study of the water system. See Section 4.8.

The results of the Spearman rank correlation will not be discussed in this example. Instead, the figures that follow, Figures 6.11 through 6.17, study water quality trends in Water System 1 using sparklines described in Section 5.7.3.

No correlations were seen between pH and alkalinity and dissolved lead and copper release. Dissolved lead and copper release trended together. See Figure 6.11.

There was water main flushing in the first August of the monitoring program near the monitoring station. The lowest ORP was seen in the system water accompanied

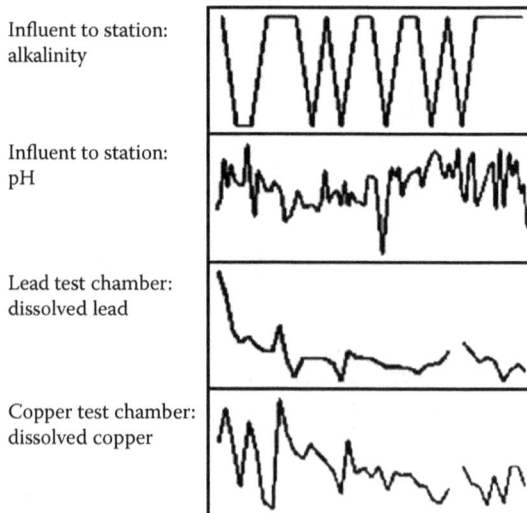

FIGURE 6.11   Water System 1: relationships of lead and copper to alkalinity and pH.

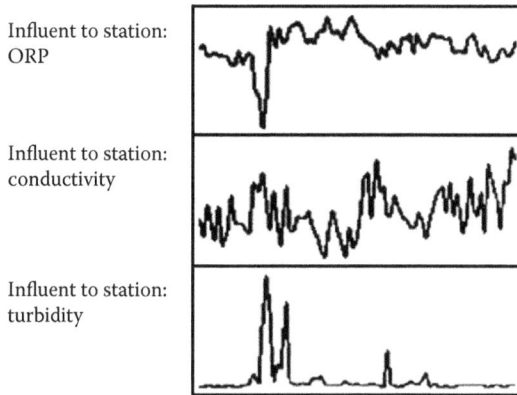

FIGURE 6.12 Water System 1: relationships of ORP, conductivity, and turbidity.

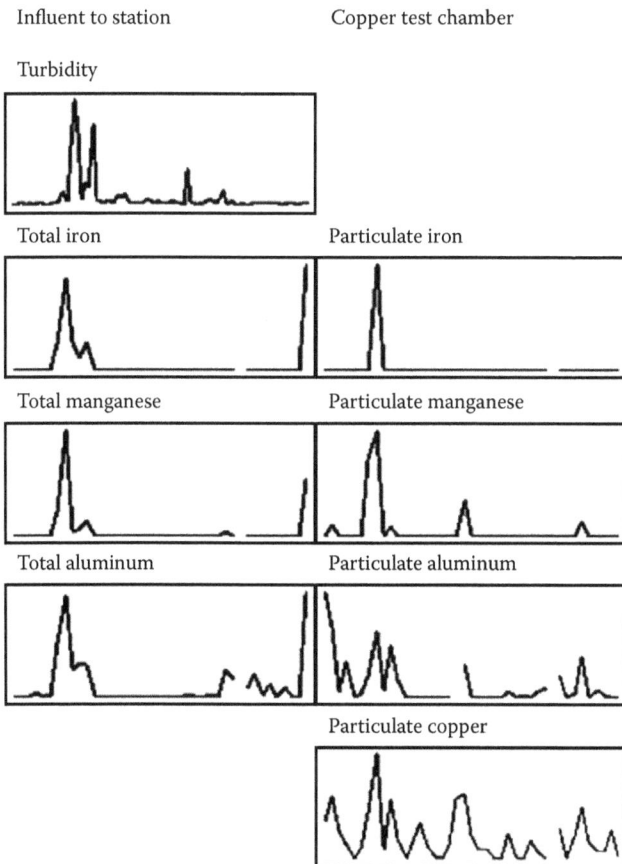

FIGURE 6.13 Water System 1: relationships of copper to other chemical scales.

Influent to station   Lead test chamber

Turbidity

Total iron   Particulate iron

Total manganese   Particulate manganese

Total aluminum   Particulate aluminum

Particulate lead

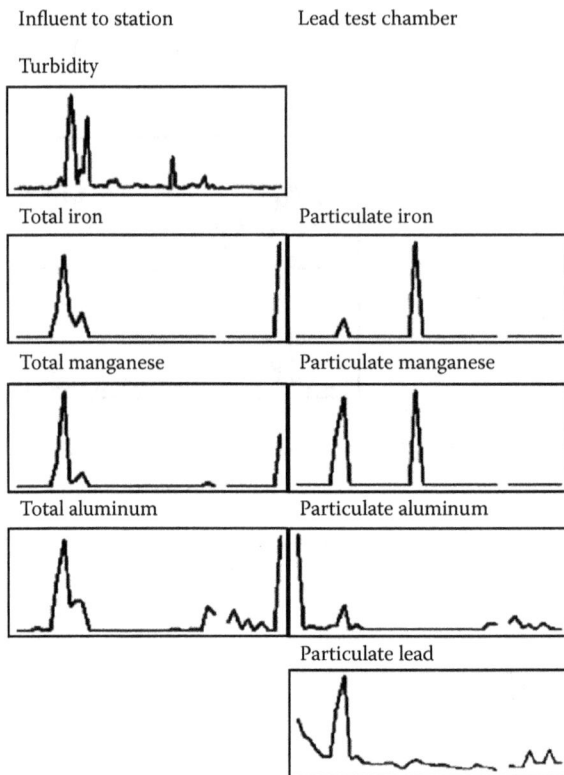

FIGURE 6.14   Water System 1: relationships of lead to other chemical scales.

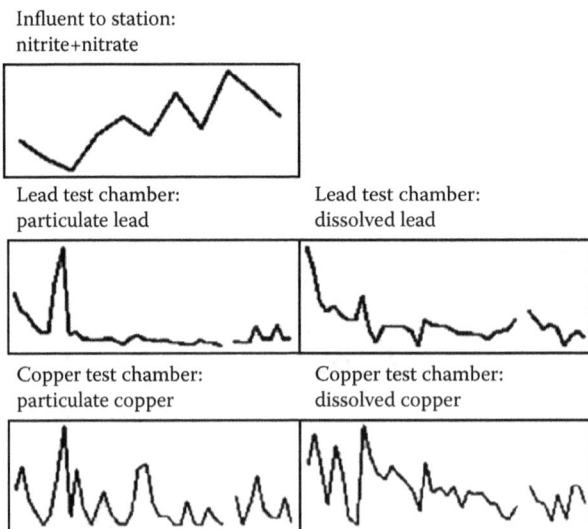

Influent to station:
nitrite+nitrate

Lead test chamber: Lead test chamber:
particulate lead   dissolved lead

Copper test chamber: Copper test chamber:
particulate copper   dissolved copper

FIGURE 6.15   Water System 1: relationships of lead and copper to nitrate concentration.

Influent to station   Test chambers

Dissolved organic carbon

Microbiological population   Lead test chamber:
                                    microbiological population

                                    Copper test chamber:
                                    microbiological population

Free chlorine

                                    Lead test chamber:
                                    dissolved lead

                                    Copper test chamber:
                                    dissolved copper

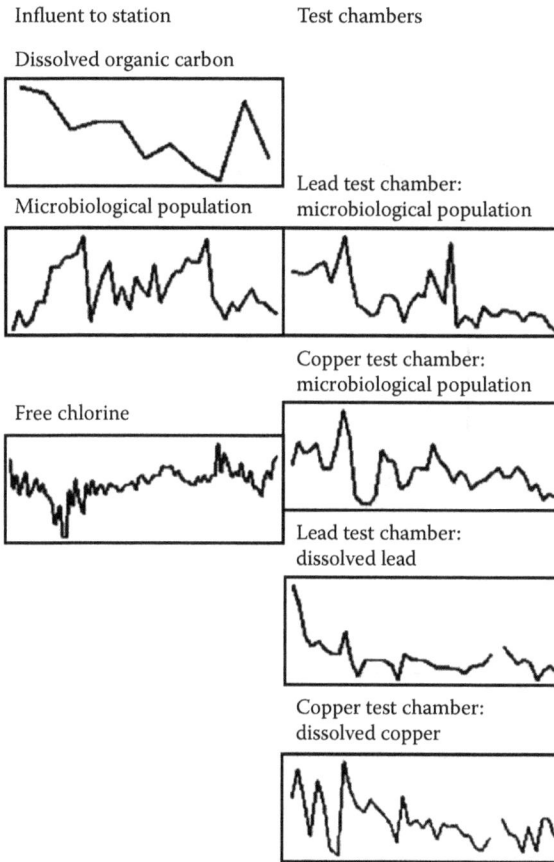

FIGURE 6.16   Water System 1: relationships of lead and copper to other biostability parameters.

by an increase in both conductivity and turbidity. Conductivity (dissolved solids) continued to increase over time. See Figure 6.12.

The August water main flushing event appeared to have increased turbidity, iron, manganese, and aluminum in the system water. The effect was seen in the release of all four metals from the copper test chamber. In general, particulate copper release trended with particulate aluminum, manganese, and possibly iron release. See Figure 6.13.

Overall, particulate lead release trended with particulate aluminum release. Particulate iron and manganese trended together and possibly with particulate lead. See Figure 6.14.

Particulate lead and copper releases trended with the increase of nitrite/nitrate in the fall (Figure 6.15). All of the metal forms decreased over time.

Figure 6.16 suggests that disinfection (free chlorine) followed an opposite trend to microbiological population. Dissolved lead and copper release appeared to follow microbiological populations in their respective test chambers.

Influent to station:
ammonia

Influent to station:
nitrite+nitrate

Influent to station:
dissolved organic carbon

Lead test chamber:
dissolved lead

Copper test chamber:
dissolved copper

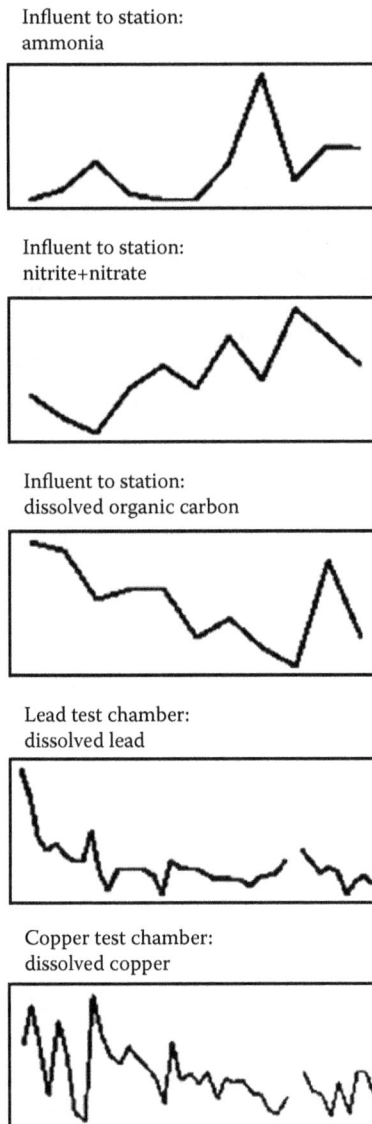

FIGURE 6.17   Water System 1: relationships of lead and copper to microbiological nutrients in the water.

The nutrient graphs of Water System 1 look similar to water system graphs where nitrification is occurring (Figure 6.17). Ammonia increased in the water system to a summertime peak. Dissolved organic carbon peaked just after the ammonia. Dissolved organic carbon appeared opposite in trend to nitrite/nitrate. Dissolved lead release, even though decreasing in general as the water system was cleaned, experienced an increase and then a decrease over the ammonia and dissolved organic carbon patterns. Dissolved copper release did this also but appeared to continue an upward climb with nitrite/nitrate, whereas dissolved lead release continued downward. These patterns are typically seen in comprehensive distribution system monitoring of chloraminated water systems undergoing seasonal nitrification. Water System 1, however, was not a chloraminated system.

It seemed odd to observe the nitrification patterns in a nonchloraminated water system and odd to see metals seemingly respond to this lower level of nitrification. A later study into the biostability of the system by others found that the 2005 installation of a second water transmission line from the lake to the treatment plant appeared to have influenced the microbiological activity in the water system. The residence time of water was greatly increased with the new transmission line and microbiological populations and biofilms increased before the water entered the treatment plant. Acetate, a simple organic carbon compound, and ammonia were found to be produced by the microbiological activity in the pipe and the specific microorganisms were identified by DNA analysis. In this way, the biostability study located the source of the ammonia, organic carbon, and nitrate patterns that were observed at the monitoring station where they were related to dissolved and particulate lead and copper release patterns.

6.1.5 Metal Plate Analyses

In addition to analyzing water samples, the metal plates from the PRS Monitoring Station were analyzed at the end of the water monitoring period. That is, at some agreed-upon period (typically at least a year, to capture seasonal trends), the water sampling program is stopped and the metal plates are removed from the test chambers for analyses in a chemical laboratory and a microbiological laboratory. Refer to Appendix A.

The chemical analysis of Water System 1 metal plate scales is shown in Table 6.5. For the lead plates, by the end of the monitoring period, 97% of the initially bare metal was covered with scales. The scales contained the common lead oxide and carbonate compounds found in surfaces exposed in water systems. There were no exceptionally protective barriers against uniform corrosion found—no highly oxidized compound called plattnerite and no orthophosphate compound called pyromorphite. (The water system does not feed a phosphate corrosion control chemical so the absence of the orthophosphate compound was not a surprise.) On the copper plates, a common copper oxide compound (cuprite) was found covering 75% of the initially bare metal. Carbonate compounds of copper, which are less soluble and more protective against uniform corrosion than cuprite, were not found. The absence of less-soluble compounds of lead and copper may be one explanation as to why the lead and copper was mostly in dissolved form in the water.

Calcium and magnesium were found in the scales, especially on the lead plates. A low but measurable amount of phosphorus was found on both the lead and copper plates even though phosphate was not added as a treatment chemical. Iron, manganese, and especially aluminum, were found on the lead and copper plates. Aluminum made up a high percentage of the chemical scales. A relationship of aluminum, iron, and manganese to particulate lead and copper might be explained by the high percentage of those metals in the plate surface scales.

TABLE 6.5

Water System 1: Metal Plate Scale Chemical Analysis

Major Minerals on PRS Monitoring Station Lead Plates by X-ray Diffraction

Scale Coverage	Litharge	Cerussite	Hydrocerussite	Plattnerite	Pyromorphite
	PbO	$PbCO_3$	$Pb_3(CO_3)_2(OH)_2$	PbO_2	$Pb_5(PO_4)_3Cl$
0.97	100	58	55		

Major Minerals on PRS Monitoring Station Copper Plates by X-ray Diffraction

Scale Coverage	Cuprite	Tenorite	Malachite
	Cu_2O	CuO	$Cu_2CO_3(OH)_2$
0.75	100		

Amounts are percent of largest X-ray diffraction peak for scale minerals
Scale coverage is the metal peak on actual plate/metal peak on pure metal

Extraneous Elements on PRS Monitoring Station Lead Plates by X-ray Fluorescence or Energy Dispersive Spectroscopy by Weight (%)

Ca	Mg	Cl
9.59	1.07	

Extraneous Elements on PRS Monitoring Station Copper Plates by X-ray Fluorescence or Energy Dispersive Spectroscopy by Weight (%)

Ca	Mg	Cl
2.96	0.94	1.04

Phosphorus on PRS Monitoring Station Lead and Copper Plates by X-ray Fluorescence or Energy Dispersive Spectroscopy by Weight (%)

Phosphorus on Lead Plates	Phosphorus on Copper Plates
0.31	0.12

Extraneous Elements on Lead Plates by X-ray Fluorescence or Energy Dispersive Spectroscopy by Weight (%)

Al	Fe	Mn
8.7	0.33	0.12

Extraneous Elements on Copper Plates by X-ray Fluorescence or Energy Dispersive Spectroscopy by Weight (%)

Al	Fe	Mn
0.23		

However, the chemical scales on the metal plate surfaces do not tell the whole story. There were also biofilms intertwined in the chemical scales. Table 6.6 shows that more biofilm formed on the copper plates than on the lead plates. But both the lead and copper test chambers had microorganisms entrained in the water and as a major presence on the metal surfaces. The presence of microorganisms and their

TABLE 6.6
Water System 1: Metal Plate Biofilm Analysis

Microbiological Population (ATP) Distributed between the Water and the Metal Surface

In System Water	On Lead Surface	In Lead Test Chamber Water	On Copper Surface	In Copper Test Chamber Water
ME/mL	ME/cm^2	ME/mL	ME/cm^2	ME/mL
2,100	3,220	128	10,200	900

Microbiological Population (ATP) Distributed between the Water and the Metal Surface (%)

On Lead Surface	In Lead Test Chamber Water	On Copper Surface	In Copper Test Chamber Water
96	4	91	9

Note: Calculations based on ATP data; metal surface area in test chamber = 854.6 cm^2; volume of water in test chamber = 950 mL.

biofilms was found to be related to dissolved lead and copper release in the water quality parameter correlations. The possible ways that microorganisms can corrode metals are described in Section 2.2.2.

6.1.6 WATER SYSTEM IMPROVEMENTS

The initial investigation theorized that legacy iron and manganese scales were responsible for release of particulate lead in Water System 1. In addition, microbiologically influenced corrosion was suspected in the corrosion of both lead and copper. A first step to remediating these issues was to carry out a high velocity unidirectional water main flushing program. The flushing was performed in the water system during the comprehensive monitoring period. A separate biostability study was performed, giving more insight into the importance of removing biofilms from the pipe walls. In the future, more work will be performed specifically on cleaning the transmission lines from the lake to the treatment plant so that microbiological products that cause microbiologically influenced corrosion are prevented from entering the distribution system.

The Lead and Copper Rule compliance data were studied to determine if progress had been made in lowering lead levels with the unidirectional flushing of the water mains. Figures 6.18 through 6.20 are box and whisker plots of Lead and Copper Rule compliance datasets. Displaying distributions of datasets were explained in Section 5.3.

Figure 6.18 shows that lead concentrations in the water system began to increase after a second transmission line was installed in 2005. The biostability study that found the addition of the second line affected water residence time also found that biofilms were responsible for an increase in dissolved organic carbon concentrations as well as ammonia, nitrates, and acetates. These chemicals can form highly soluble

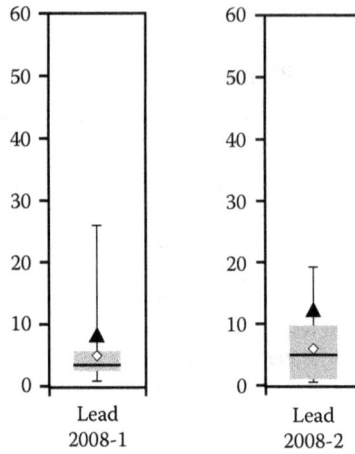

FIGURE 6.18   Water System 1: Lead and Copper Rule lead data after the 2005 installation of a new transmission line (lead in µg/L).

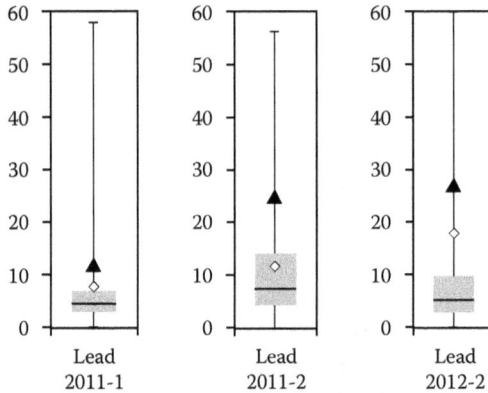

FIGURE 6.19   Water System 1: Lead and Copper Rule lead data after 2010 water system repairs to the older transmission line (lead in µg/L).

compounds with lead and copper. Their relationships to Water System 1 dissolved lead and copper release were seen in the PRS Monitoring Station data (Figure 6.17).

Figure 6.19 shows that lead concentrations in the water system made a bigger jump after repairs to the water system occurred in 2010 and the older transmission line in 2011. The biostability study found that construction in the existing transmission lines and the treatment plant released high concentrations of dissolved organic carbons from existing biofilms into the water system. This provided additional pathways to lead and copper corrosion and the tendency of the water in the distribution system to form biofilms, causing the perpetuation of the production of microbiological waste products that were soluble to lead and copper.

Figure 6.20 shows the decrease of lead concentrations after unidirectional flushing of water mains in 2014 and 2015. Unfortunately, there was no flushing in 2016, so

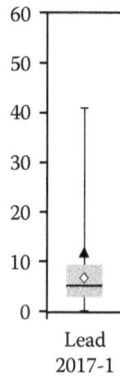

FIGURE 6.20   Water System 1: Lead and Copper Rule lead data after 2014 and 2015 unidirectional water main flushing.

the cleaning process was disrupted. The 90th percentile lead concentration hovered just above the Action Level of 15 µg/L in the 2016-2 compliance sampling event. The cleaning efforts were restarted and 2017 compliance sampling showed the 90th percentile at 12 µg/L. The 2017 summer cleaning efforts were expected to continue lead compliance and possibly move the 90th percentile concentration further below the Action Level. One more required sampling event in summer 2018 is expected to put the water system back into total compliance with the Lead and Copper Rule.

6.1.7 CONCLUSIONS

The monitoring data show that the water system cleaning was successful in improving the water quality of Water System 1. Further recommendations for this water system were as follows:

- Plan for the removal of complete lead service lines and galvanized iron service lines.
- Continue to make water main high velocity cleaning a priority. Aluminum from coagulant and iron and manganese from intermittent use of groundwater must be routinely cleaned away from pipe walls. The scale is the precursor to transport of particulate lead and copper in premise plumbing. High velocity flushing also removes biofilms, which is essential to controlling a significant factor of lead and copper release in this water system.
- Plan for the replacement of older and corroded water main.
- Continue on the path to controlling the biostability of the transmission line water.
- Continue on the path to controlling the biostability of the water filter.
- Plan for investigation and routine maintenance of the standby wells for biostability.
- Consider incorporating the use of the wells into a more routine pumping/water blending program in order to achieve better biostability in the wells and to achieve more uniform chemical characteristics of system water.

6.2   WATER SYSTEM 2

Water System 2 has utilized PRS Monitoring Stations over several time periods. The first time the monitoring stations were used in the water system was to prepare for a major operational change of disinfection chemical from free chlorine to chloramine and a possible change of corrosion control chemical. The monitoring started with one monitoring station at the entry point to the distribution system in order to define the characteristics of the water system in its original configuration using free chlorine disinfection and a 50% polyphosphate/50% orthophosphate corrosion control chemical. The PRS monitoring station also tracked the lead and copper release and microbiological growth in the test chambers.

A second monitoring station was set up at the same time in an off-line chemical comparison test using lead test chambers. Treatment plant water was prepared with chloramine disinfection and no corrosion control chemical, which served as the influent water to the monitoring station and one test chamber. Three other lead test chambers in the monitoring station also received the chloraminated water. Each of those test chambers were also dosed with corrosion control products—one chamber was dosed with 50% polyphosphate/50% orthophosphate, which was being used in the water system at the time; a second chamber was dosed with 70% polyphosphate/30% orthophosphate; a third chamber was dosed with 10% polyphosphate/90% orthophosphate. The product with the highest orthophosphate was found to release the least total and dissolved lead into the stagnating water of a lead test chamber, as shown in Figure 6.21.

With the results of the chemical comparison tests, it was decided to switch the water system's corrosion control product from a 50% polyphosphate/50% orthophosphate chemical to a 10% polyphosphate/90% orthophosphate chemical.

At the end of the chemical comparison tests, the PRS Monitoring Station was moved to a high water age location in the distribution system to define the characteristics of the water system in its original configuration using free chlorine disinfection and a 50%

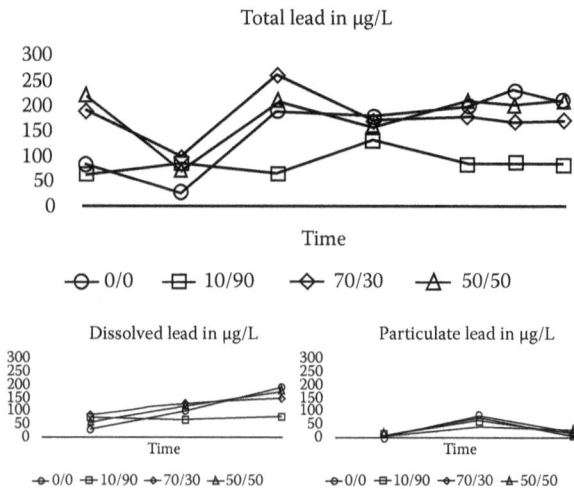

FIGURE 6.21   Water System 2: corrosion control chemical comparison tests using a PRS Monitoring Station.

polyphosphate/50% orthophosphate corrosion control chemical at that location. The PRS monitoring station also tracked the lead and copper release and microbiological growth in the test chambers, which were compared to the results at the distribution system entry point PRS Monitoring Station.

Both monitoring stations were operating when the corrosion control chemical was changed and, two months later, the disinfection chemical was changed. More frequent sampling during the chemical transition gave confidence that lead release was being lowered as predicted by the off-line chemical comparison tests.

Monitoring continued for several months after the chemical transition to track the lead and copper release and other water quality characteristics.

Several more months after the end of the chemical transition monitoring, the two PRS Monitoring Stations were restarted for a year of monitoring (November 2009 through November 2010) to better understand the factors shaping the water quality under the new treatment conditions and to look for opportunities for ongoing water quality control and system improvement. That monitoring confirmed that a significant presence of aluminum developed in the metal plate accumulations. The aluminum appeared to be related to particulate lead release. The development of biofilms in the metal plate accumulations and patterns of nitrification in the chloraminated system were also evident. A focus was placed on the need to clean pipe wall chemical scale and biofilm accumulations out of the water system using high velocity unidirectional flushing. Money and labor were not available for that action.

Four years after that study (2014), monitoring with one PRS Monitoring Station at a high water age location began again for more than one year. The findings of significant particulate lead release, a high and possibly influencing presence of aluminum, and a nitrification cycle that corresponded to particulate lead and copper and dissolved lead and copper release were confirmed and studied in greater detail.

6.2.1 COMPARISON OF PRS MONITORING STATION DATA TO RESIDENTIAL DATA

During the PRS Monitoring Station studies, residences with lead service lines were sampled at times to compare to PRS Monitoring Station test chamber data. During the 2009/2010 study, residences were sampled in March, July, and October, with first-draw stagnation samples similar to Lead and Copper Rule compliance sampling. The PRS Monitoring Station lead and copper release results were higher than the residential data as expected, given the exposure to pure lead in the lead test chamber and pure copper in the copper test chamber, the younger scale development in the test chambers, and the exaggerated stagnation period in the monitoring station. The lead test chamber had total lead concentrations of 60 to 100 µg/L in stagnating water, while the residences had first-draw lead concentration ranges of around 5 to 15 µg/L. The copper test chamber had total copper concentrations of 125 to 200 µg/L, while the residences had copper concentration ranges of around 10 to 200 µg/L. The copper concentrations in the test chambers were closer to residential magnitudes because residences typically have copper piping within the first liter of water withdrawn from a kitchen sink faucet—a typical sampling location for residences.

A better view of lead concentrations in lead service lines and not just at the kitchen faucet was obtained in the 2014 monitoring period. As described for Water System 1 and in Section 4.7, profile sampling was performed on two residences. Figure 6.22 shows the results of the residential sampling. Lead concentrations were higher in the lead service lines than in the first liter taken from the kitchen faucet. Lead release was the highest in the November sampling versus the earlier sampling. Figure 6.23 shows that more than 50% of the total lead in the Water System 2 residences was in particulate form.

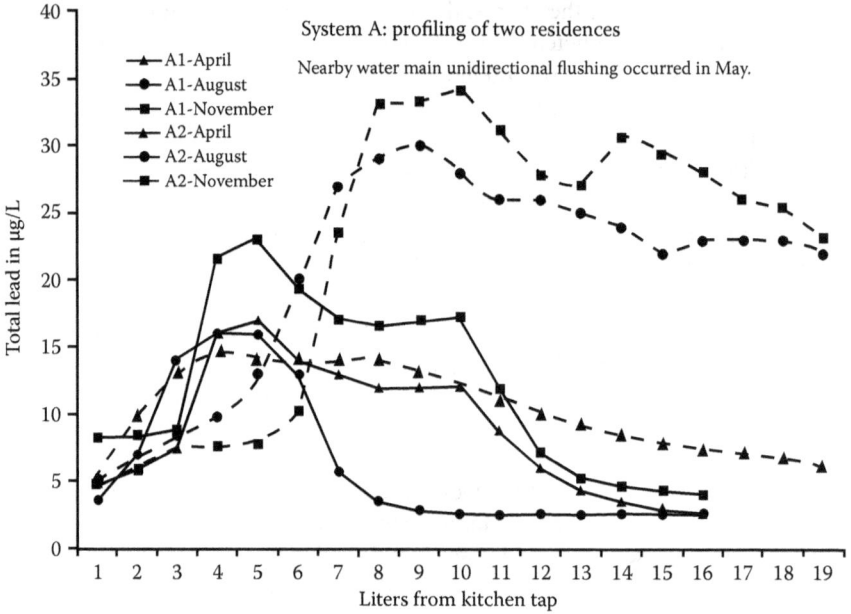

Site A1		Site A2	
Liters from kitchen tap	Piping material	Liters from kitchen tap	Piping material
1	Copper	1	Copper/galvanized
2	Copper	2	Copper
3	Galvanized iron	3	Copper
4	Galvanized iron	4	Copper
5	Lead	5	Copper
6	Lead	6	Lead
7	Lead	7	Lead
8	Lead	8	Lead
9	Lead	9	Lead
10	Lead	10	Lead
11	Lead	11	Lead
12	Lead	12	Lead
13	Lead	13	Lead
14	Lead	14	Lead
15	Water main	15	Copper
16	Water main	16	Copper
		17	Copper
		18	Water main

FIGURE 6.22 Water System 2: residential profile sampling total lead concentrations.

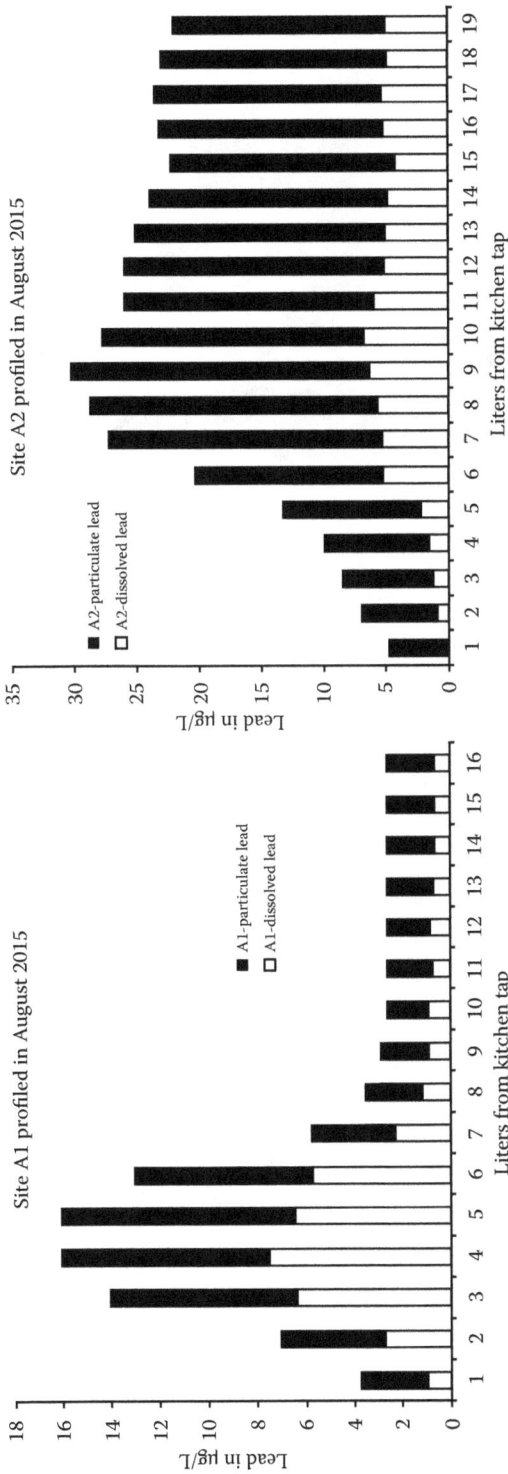

FIGURE 6.23 Water System 2: residential profile sampling dissolved and particulate lead concentrations.

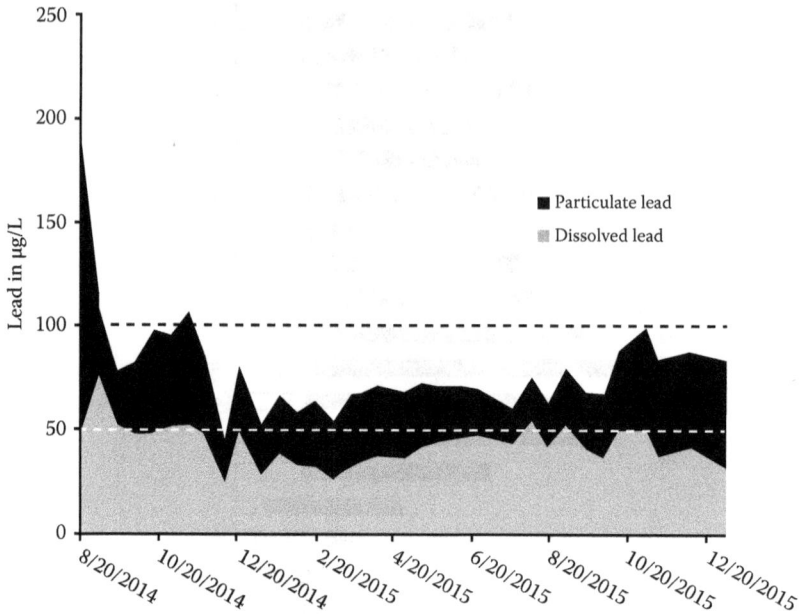

FIGURE 6.24   Water System 2: lead release in the PRS Monitoring Station lead test chamber.

Figure 6.24 shows that the PRS Monitoring Station lead test chamber predicted the high percentage of particulate lead release. It also predicted higher release of particulate lead in November.

A dotted horizontal line on the graph in Figure 6.24 at 100 μg/L is an empirical guideline from PRS Monitoring Station results. It has been observed that water systems with lead data under 100 μg/L in a PRS Monitoring Station lead test chamber are typically in compliance with the Lead and Copper Rule's first-liter lead concentration criteria. With lead test chamber data under 50 μg/L, lead release is anecdotally predicted to be in control, even in lead service lines where a source of pure lead is found in a water system. In Water System 2, the dissolved lead release in the lead test chamber presented the possibility of excellent lead release control in the actual water system. The total lead concentration, where the particulate lead fraction is added to the dissolved lead levels, was predicted to be in compliance with the Lead and Copper Rule but with the possibility of higher values occurring in lead service lines, especially in the autumn when the particulate lead release peaked. This was confirmed in the residential sampling.

6.2.2   COMPARISON OF PRS MONITORING STATION DATA TO LEAD AND COPPER RULE DATA

The study of Water System 2 with PRS Monitoring Stations began in 2008, as described previously. The lowering of lead concentrations with a change of corrosion control chemical was predicted by monitoring data. The 2009 Lead and Copper

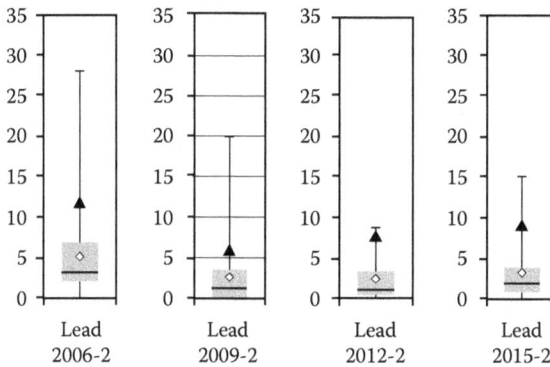

FIGURE 6.25   Water System 2: Lead and Copper Rule lead data box and whisker plots.

Rule data shown in Figure 6.25 shows lower median, average, 90th percentile and maximum lead concentrations after the corrosion control chemical change.

The 2015 Lead and Copper Rule data in Figure 6.25 hints that lead concentration levels could possibly be on the increase. The 2014/2015 PRS Monitoring Station study indicated continued issues with aluminum accumulations, biofilms, and nitrification patterns that trended with lead release. The need for efficient flushing of pipe wall accumulations and for better nitrification control appears to be more pressing for Water System 2 given the patterns seen in the Lead and Copper Rule data.

6.2.3   Conclusions

Water System 2 has used PRS Monitoring Stations and the comprehensive approach to water quality control for various reasons—defining the characteristics of a water system, performing off-line chemical comparison tests, looking for simultaneous regulatory compliance while making treatment chemical changes, and monitoring routinely for water quality control and process improvement. The monitoring data provided guidance for operational decisions. Some of the guidance had not yet been followed and it is predicted that increasing lead levels in the water may result.

6.3   WATER SYSTEM 3

Comprehensive monitoring in Water System 3 using two PRS Monitoring Stations, both at high water age locations in the distribution system, was performed over several water treatment changes. The data showed trends in water quality parameters with the water treatment changes including lead and copper release trends and microbiological growth patterns.

Regarding lead and copper release in the PRS Monitoring Stations' test chambers, the metals concentrations, especially in particulate form, trended with sporadic events in the distribution system. The events were characterized by sudden high concentrations of iron particulates and microbiological populations in the flowing system water entering the PRS Monitoring Station. This appeared to be related to release of chemical scales and biofilm accumulations from the pipe walls of corroded unlined cast iron water main that was predominant in the distribution system.

After high-velocity unidirectional flushing, there were fewer releases of lead and copper and concentrations of released metals were lower overall.

6.3.1   COMPARISON OF PRS MONITORING STATION DATA TO ROUTINE BUILDING SAMPLING

During this monitoring period, a commercial building was sampled weekly along with the two PRS Monitoring Stations. Water utility personnel were given access to the building early in the morning before the building was occupied so that a first-draw stagnation sample could be obtained from a designated faucet. After the stagnation sample was obtained, the water was allowed to flow and a flowing water sample was obtained similar to the flowing system water sample taken at the influent sample tap of the PRS Monitoring Station.

Figure 6.26 compares total lead release from lead test chambers at the two PRS Monitoring Station locations. They were similar in magnitude of lead concentration except for the timing of the localized high-lead events. The figure also displays the similarities of the total lead release from a high-leaded brass test chamber to the lead release in the commercial building samples. Because the erratic high-lead events were mostly of particulate lead (about 78%), the dissolved lead release between the pairs of sites was more similar than the total lead release, as shown in Figure 6.27.

The trends of copper from the sampling sites (a copper test chamber, a brass test chamber, and the commercial building) were similar to that of lead, showing sporadic peaks in concentration. Low levels of copper were mostly in dissolved form, but the high peaks were about 93% particulate copper. Also, the dissolved copper released from the brass test chamber and from the commercial building were similar in magnitude and trends.

6.3.2   COMPARISON OF PRS MONITORING STATION DATA TO RESIDENTIAL DATA

Over the same 2013/2014 time period that the PRS Monitoring Stations were sampled, profile sampling of eight residences was performed. Then, samples were obtained monthly only from within the lead service lines in order to track lead release from a location with the highest release potential.

Tables 6.7 and 6.8 show a comparison between metal release results from the PRS Monitoring Stations' copper and brass test chambers, the commercial building, eight lead service lines, and eight first-draw kitchen faucets. In these tables, dissolved copper, total copper, dissolved lead, and total lead are studied.

In Table 6.7, the highest expected value at each site, a calculated statistic, is compared. No data are listed if the metal concentration was significantly lower than the highest results. The data in the table show that dissolved copper release in the copper test chamber was higher than residential data but of the same magnitude. Dissolved copper release from the brass test chamber was similar to the highest dissolved copper values measured in the residential lead service lines and at the kitchen faucets.

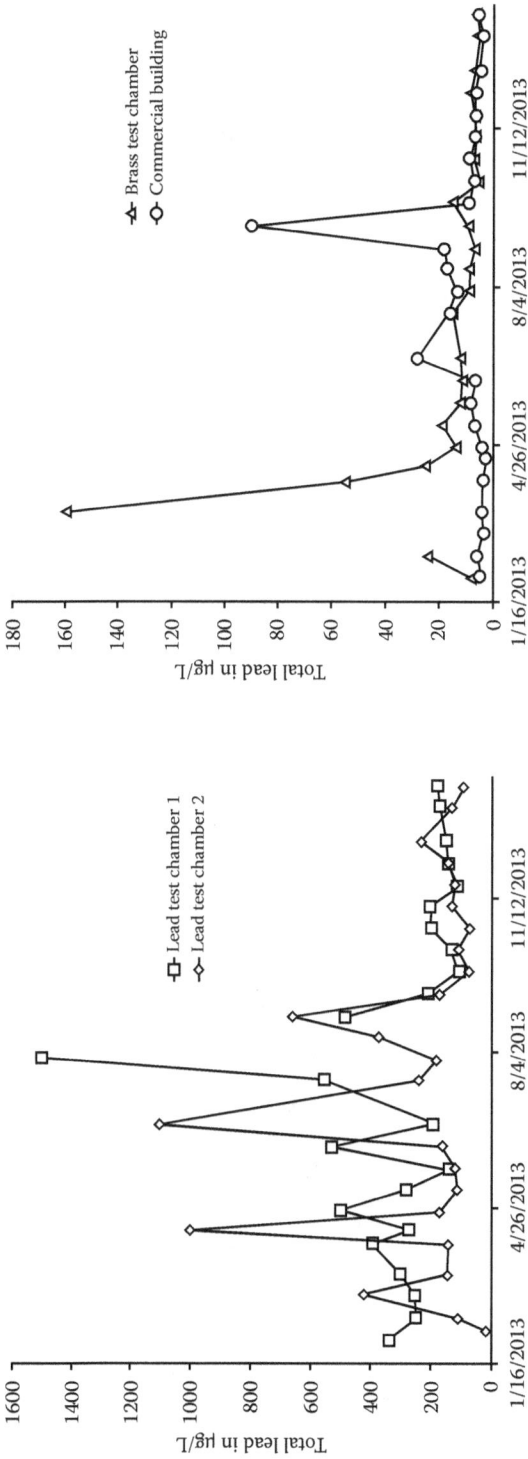

FIGURE 6.26 Water System 3: total lead release at two PRS Monitoring Stations and a commercial building faucet.

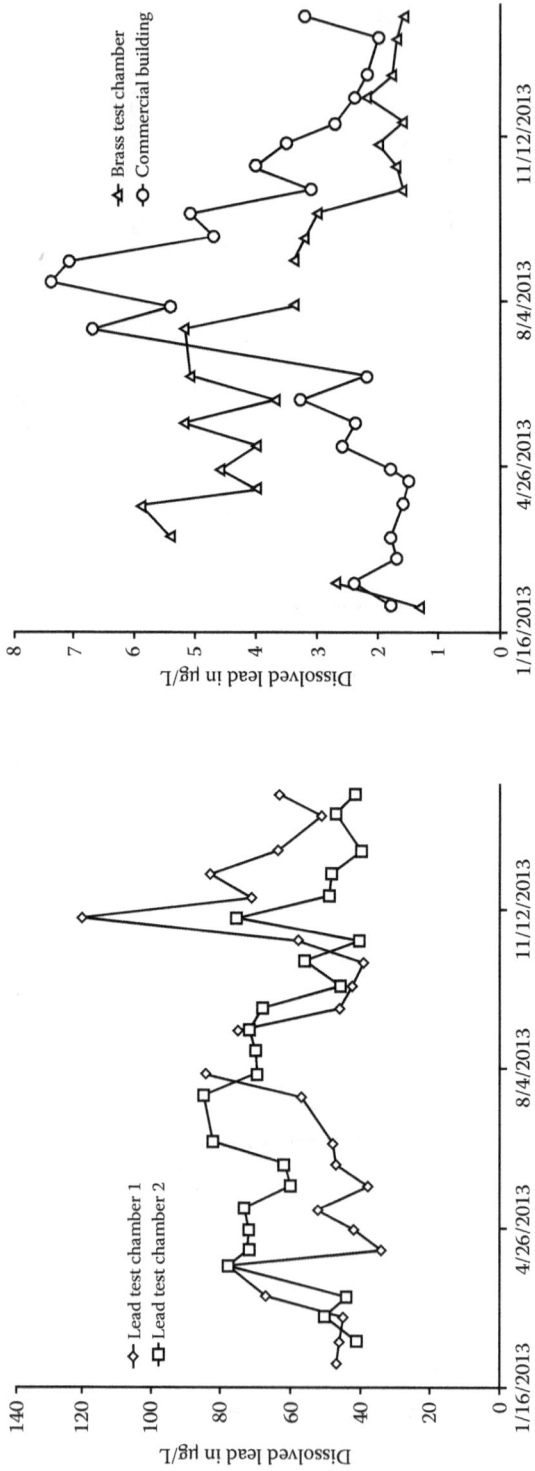

FIGURE 6.27   Water System 3: dissolved lead release at two PRS Monitoring Stations and a commercial building faucet.

TABLE 6.7

Water System 3: Comparison of Test Chamber Metal Release to Residential Metal Release—Highest Expected Values

Site	Dissolved Copper (µg/L)	Total Copper (µg/L)	Dissolved Lead (µg/L)	Total Lead (µg/L)
Copper test chamber	74	2800	–	–
Brass test chamber	38	920	–	–
Lead test chamber 1	–	–	100	750
Lead test chamber 2	–	–	90	890
Commercial building	–	–	–	–
Lead service line	39	–	76	138
First draw at kitchen sink	37	161	30	85

TABLE 6.8

Water System 3: Comparison of Test Chamber Metal Release to Residential Metal Release—Average Values

Site	Dissolved Copper (µg/L)	Total Copper (µg/L)	Dissolved Lead (µg/L)	Total Lead (µg/L)
Copper test chamber	26	540	–	–
Brass test chamber	17	180	–	–
Lead test chamber 1	–	–	58	320
Lead test chamber 2	–	–	58	250
Commercial building	11	19	–	–
Lead service line	8	6	30	65
First draw at kitchen sink	18	64	–	–

Total copper release in the test chambers was many magnitudes higher than residential data. This may have been because of the direct exposure of the test chambers to pipe wall accumulation events in the distribution system.

Dissolved lead release from the lead test chambers was higher but of the same magnitude as measured in lead service lines and at kitchen faucets. Total lead release in the lead test chambers was about eight times higher than residential release. Similar to the total copper release, the total lead release signified particulate pipe wall accumulation release that was more localized.

Similar to the highest expected values of Table 6.7, the compared sampling data averages in Table 6.8 show that the dissolved lead and copper data between test chambers and residences were within the same magnitude. The total lead and copper data, which included particulate lead and copper release, were more dependent on locality in the receipt of particulate pipe wall accumulations, which appeared to influence particulate lead and copper release.

6.3.3 COMPARISON OF **PRS** MONITORING STATION DATA TO LEAD AND COPPER RULE COMPLIANCE DATA

Statistics calculated from Lead and Copper Rule data from Water System 3 (maximum concentration, 90th percentile concentration, and average concentration) were plotted in comparison to the PRS Monitoring Stations' and commercial building's data. The Lead and Copper Rule statistics were calculated from about 100 to 150 residential water samples; the Monitoring Station and commercial building data were from the individual locations only, so more variability in the data was expected than for the Lead and Copper Rule statistics. In addition, the high erratic jumps of particulate lead and copper measured in the test chambers of Water System 3 appeared to have been damped down in magnitude within the building plumbing systems, so the same lower concentrations would be expected in Lead and Copper Rule data that come from buildings.

The maximum of the Lead and Copper Rule (LCR) datasets over time were plotted against data from the PRS Monitoring Station copper test chamber, as shown in Figure 6.28. The LCR maximum values were similar to the lower concentrations of the copper test chamber. That is, the copper test chamber experienced the high erratic jumps in copper, as previously described, which were not captured by the LCR data.

In Figure 6.29, the 90th percentile LCR copper concentration data were similar to the lower curve of the PRS Monitoring Station brass test chamber. Again, erratic

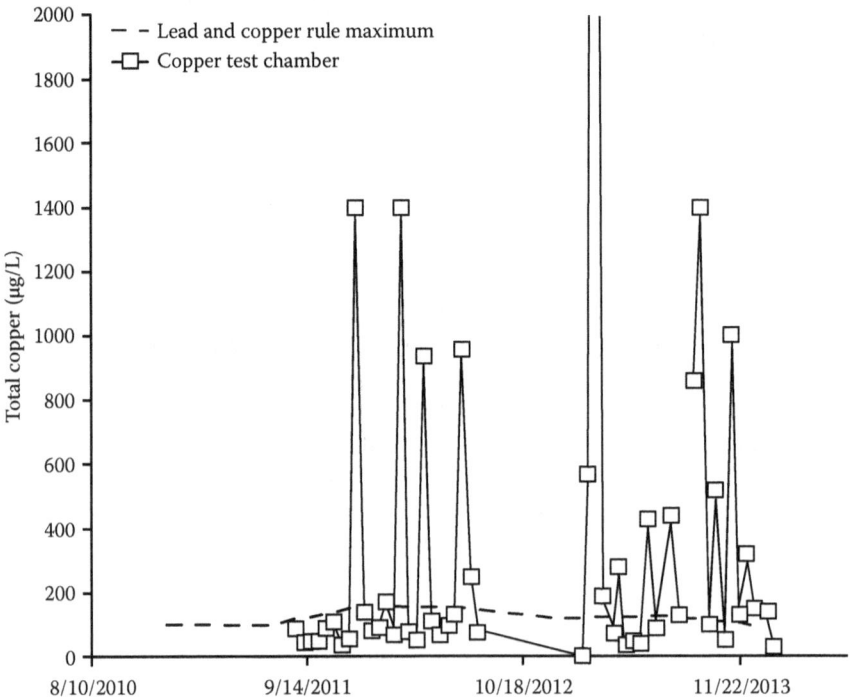

FIGURE 6.28 Lead and Copper Rule maximum copper concentrations versus PRS Monitoring Station copper test chamber data.

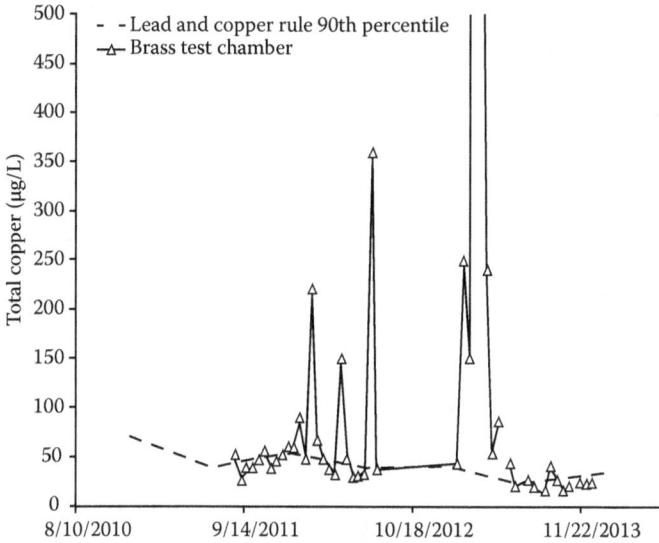

FIGURE 6.29 Lead and Copper Rule 90th percentile copper concentrations versus PRS Monitoring Station brass test chamber data.

jumps in concentration occurred in the test chamber data but not the LCR dataset statistics.

In Figure 6.30, the average copper concentration of the LCR datasets over time was plotted against data from the commercial building. The average concentrations

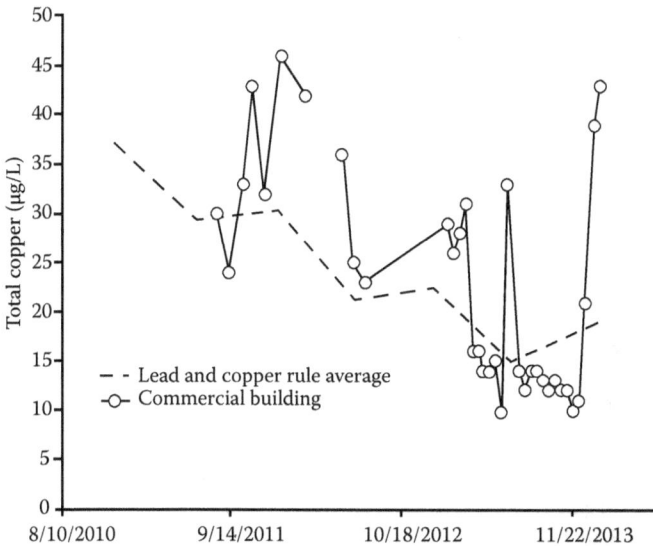

FIGURE 6.30 Lead and Copper Rule average copper concentrations versus commercial building data.

were similar to the trend of copper release in the commercial building, where jumps in concentration are present but not of the magnitude seen in the test chambers.

The same relationships applied to LCR data lead statistics: Maximum LCR lead concentrations equaled PRS Monitoring Station lead test chamber data;

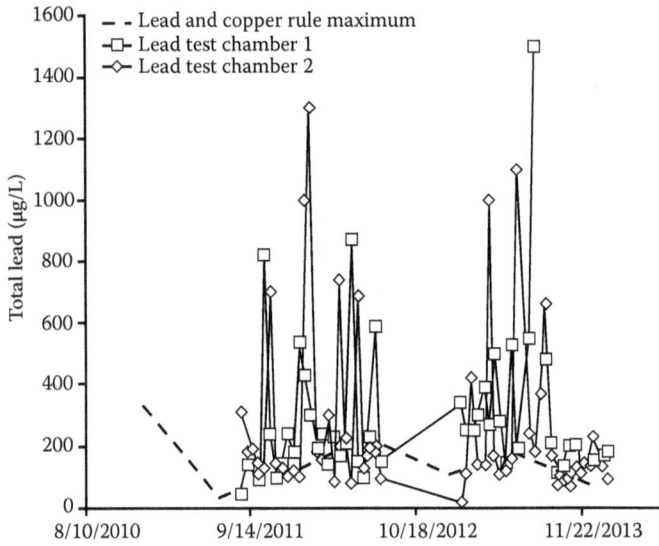

FIGURE 6.31 Lead and Copper Rule maximum lead concentrations versus PRS Monitoring Station lead test chamber data.

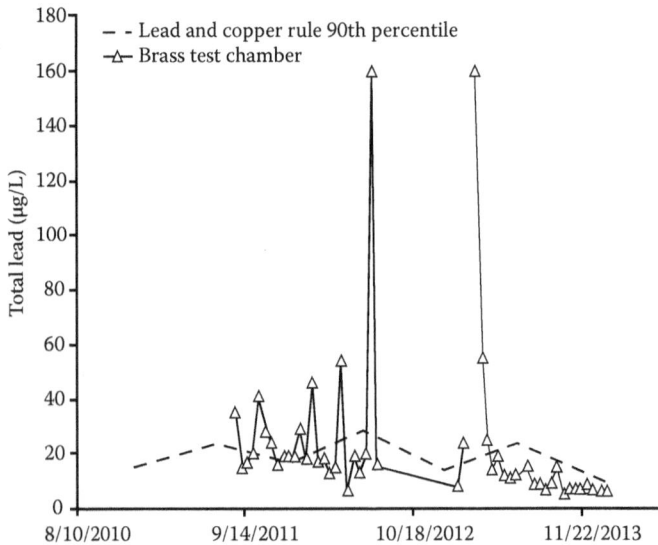

FIGURE 6.32 Lead and Copper Rule 90th percentile lead concentrations versus PRS Monitoring Station brass test chamber data.

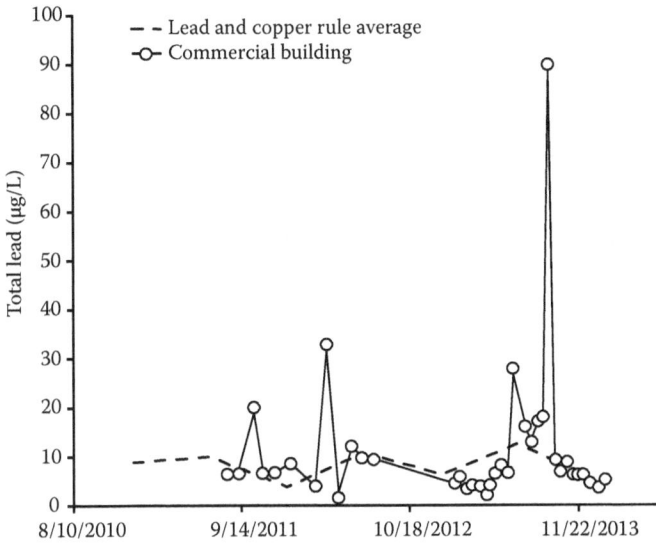

FIGURE 6.33   Lead and Copper Rule average lead concentrations versus commercial building data.

90th percentile LCR lead concentrations equaled brass test chamber data, average LCR lead concentrations equaled commercial building data (Figures 6.31 through 6.33).

6.3.4   CONCLUSIONS

PRS Monitoring Stations and a comprehensive monitoring strategy were used in Water System 3. The monitoring data identified the effects of water treatment changes and defined possible reasons that lead and copper were released into the water system.

The data appeared to be representative of residential sampling where dissolved copper release from a PRS Monitoring Station copper test chamber was found to be similar to first-draw sampling at a kitchen sink, and dissolved lead release from a PRS Monitoring Station lead test chamber was found to be similar to sampling within a lead service line.

Particulate lead and copper release differed in magnitude and timing from residential release data. Particulate lead and copper release was a function of release of pipe wall accumulations, which was a local phenomenon in Water System 3. Nevertheless, the relationship of lead and copper release to release of pipe wall accumulations was demonstrated and emphasized the need to remove pipe wall chemical scales and biofilms from the water system.

Relationships were also found between the PRS Monitoring Station data and Lead and Copper Rule compliance data.

7 A Philosophy of Drinking Water Quality Control

In this book, the control of water quality in drinking water systems is based on a comprehensive perspective. That is, water quality is shaped by complex interactions of drinking water, which is a potpourri of natural and treatment chemicals and naturally occurring microorganisms, with pipe wall accumulations of various chemical scales and biofilms. No scientific formula can predict the characteristics of the final water quality. All distribution system water quality issues (discolored water, disinfection by-products, presence of pathogens, release of lead and copper, etc.) are inter-related; they are all manifestations of the complex interactions between a complex solution of water and a complex composition of pipe wall accumulations.

To put the comprehensive perspective into action and to control water quality in distribution systems:

- Keep the pipes, tanks, filters, and wells clean of chemical scales and biofilm accumulations.
- Keep the water biologically stable by limiting nutrients in the water (organic carbon, nitrogen, and phosphorus), limiting residence time of water in the system, and providing adequate disinfection.
- Routinely monitor the water quality in the distribution system.

This book was written to guide water utility personnel in effectively monitoring the distribution system as efforts are made to understand the key factors that shape the water quality in a specific system and to improve the water quality. Effective monitoring also leads to timely responses to issues should the water quality begin to degrade.

While this proactive monitoring costs money, it appears to pay off in the long run by helping to

- Prevent regulatory compliance problems
- Eliminate unnecessary chemical costs
- Guide operational decision-making
- Increase consumer confidence

The effective monitoring protocols described in this book have been borrowed from industrial quality control and process improvement. They are summarized in the flow diagram of Figure 4.5, "Path to Control of Drinking Water Quality." The protocols include strategic planning, initial monitoring, and hypothesis testing and remediation. The final step is routine monitoring and system improvement where water quality goals have been achieved and effort can be spent zeroing in on stricter water quality parameter average values and on narrowing variation from the average.

What more can be said to encourage this effort toward high drinking water quality? Perhaps we can call on W. Edwards Deming, the industrial process improvement consultant, to remind us that a graph of a water quality parameter over time is "the process talking to us" (Deming 1993).

It is time that we listened.

Appendix A: Process Research Solutions Monitoring Station

The Process Research Solutions (PRS) Monitoring Station is intended to operate like a standard or mini pipe loop apparatus (Section 4.2.1). That is, the device is intended to provide controlled contact between distribution system water and metal in a standardized configuration. The PRS Monitoring Station always has the same surface area of desired metal-to-volume-of-water ratio in every installation. Pipe loop apparatuses, however, will vary from installation to installation based on the types and sizes of pipes installed.

The PRS Monitoring Station, similar to pipe loop apparatuses, is an idealization of a building plumbing system. The water flowing into the device is controlled by a timer that opens and closes a valve at controlled intervals. The water flows through the apparatus once and is disposed of down a wastewater drain. This once-through intermittent flow is similar to water flow in a building. The advantage of the monitoring station or pipe loop apparatus is that water usage and flow rate can be controlled and repeated for all sampling events. For the PRS Monitoring Station, a set flow rate and water usage pattern is recommended so that data from one monitoring period or water system can be compared to data from other monitoring periods or water systems.

Another difference between a pipe loop and the PRS Monitoring Station is the configuration of the metal in contact with the water. Actual pipe takes up space and is unwieldy to install. The PRS Monitoring Station eliminates the pipe altogether. Instead, 16 square metal plates (2 inches × 2 inches × 1/16 inch or 5.08 cm × 5.08 cm × 0.159 cm) are stacked inside a 4-inch (10.16-cm) diameter plastic pipe, 6.75 inches (17.14 cm) long. These measurements create a test chamber that holds a certain volume of water in contact with a certain surface area of metal. It is the same ratio of water volume to metal surface area that is found inside a 1.77-inch (4.50-cm) diameter metal pipe. It is also the most practical water-volume-to-metal-surface-area ratio to use in selecting plate and test chamber size.

A major benefit of using metal plates exposed to water is that the plates serve as excellent specimens for the study of pipe films and scales. The analysis of accumulations on the pipe walls is essential for understanding the chemical and microbiological mechanisms occurring in the water. (Analyses are described in Appendix B. This information can indicate how older pipes in the distribution system most likely interact with the water. Routine water sample data and the final metal plate film analysis data from the PRS Monitoring Station support a comprehensive picture of the water quality in the distribution system.)

Because of its configuration, the PRS Monitoring Station is relatively compact in design compared to a standard pipe loop apparatus. Figure A.1 shows various configurations of the PRS Monitoring Station over the years.

(a)

(b)

(c)

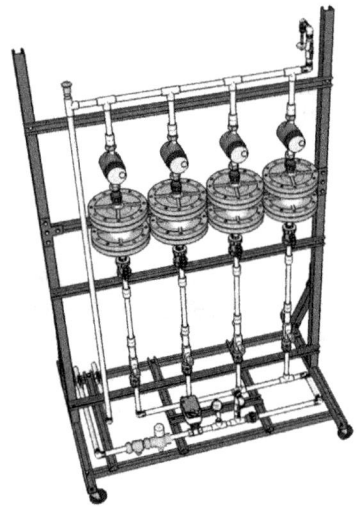

(d)

FIGURE A.1 Configurations of the PRS Monitoring Station. (a) This was the first PRS Monitoring Station. It was assembled by personnel at the Waukesha Water Utility in Wisconsin in 2006. (b) The second iteration of the PRS Monitoring Station was assembled for North Shore Water Commission in Wisconsin in 2008 by Aqualogix, Inc. (c) The Rundle-Spence company of New Berlin and Madison, Wisconsin made design improvements with PRS and began to assemble PRS Monitoring Stations in 2009. (d) In 2015, a PRS Monitoring Station with four test chambers was added to Rundle-Spence assemblies.

A.1 ASSEMBLY

A diagram of a PRS Monitoring Station is shown in Figure A.2. Components are numbered and described in Table A.1.

In general, all wetted parts of the monitoring station should be plastic except for the metal plates in the test chambers. Testing of metals in the water is an important aspect of using the monitoring station, so the surface area of metals in contact with the water should be known and controlled by eliminating all other sources of metals. Metal can be used, if necessary, upstream of the influent sample tap, since the concentration of metals influent to the test chambers are determined at that point.

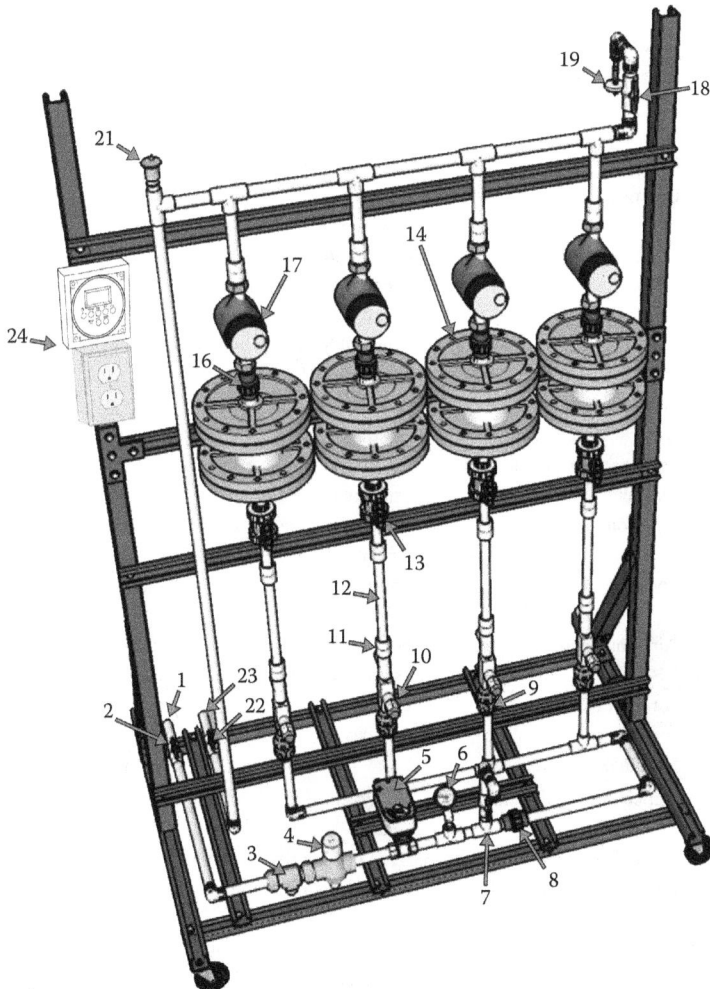

FIGURE A.2 PRS Monitoring Station with components numbered and described in Table A.1.

TABLE A.1
PRS Monitoring Station Components Description for Figure A.2

Diagram No.	Equipment	Comment
1	Pipe connection, ½" dia., female-threaded	For connecting to the water source
2	Manual influent ball valve	Keep open during operation of station
3	Vented double-check valve	For backflow prevention
4	Pressure-reducing valve	Operate station at a pressure between 15 and 50 psig; 30 psig is suggested
5	Electrically actuated influent ball valve	Controlled automatically by timer to start and stop flow to the station
6	Pressure gauge	
7	Influent sample tap	For sampling water flowing into the station
8	Pipe coupling	For removing the influent line equipment for service, if needed
9	Check valve	To prevent mixing of water from test scenarios
10	Needle valve	To set flow rate for test chamber
11	Chemical injection port, ½" dia., female-threaded	Insertion port for chemical injector, if adding chemicals to test scenario
12	Static mixer	For mixing of influent water and chemicals, if adding chemicals to test scenario
13	Union ball valve	To isolate and remove complete test chamber from vertical line, for maintenance; keep open during operation of station
14	Test chamber	Holds 2 stacks of metal plates for exposure to water for testing
15	Test chamber sample tap (not seen on drawing)	From bottom of test chamber for sampling water that has stagnated in contact with the metal plates
16	Pipe coupling	Aids in opening the top of the test chamber for installing and removing the internal metal plates
17	Totalizing flow meter	For monitoring flow rate and total flow in the test chamber line
18	Air vent manual ball valve	For allowing air into piping to push the stagnation water sample out of the test chamber sample tap
19	Air filter cartridge	For filtering microorganisms and debris out of the air that pushes out the stagnation water sample
20	left blank intentionally	
21	Antisiphoning air vent	To prevent siphoning from effluent drain line back into the station
22	Manual effluent ball valve	Keep open during operation of station
23	Pipe connection, ½" dia., female-threaded	For connecting the effluent line to a drain
24	Timer with connected duplex outlet	For automatic control of water flow through the station with automatic control of equipment, typically chemical feed pumps, plugged into the connected electrical outlet

All materials in the monitoring station should be tolerant of a 50 mg/L chlorine solution with exposure of 24 hours for cleaning and disinfecting the station before use for monitoring.

Schedule 80 plastic piping is used in the assembly, not for pressure requirements, but for structural stability.

A.1.1 INFLUENT LINE

The influent line to the PRS Monitoring Station begins with a ½-inch diameter female-threaded pipe connection. The pipe connection gives the flexibility to connect to the water source by means of plastic piping, tubing, or a hose using a threaded adapter in the connection as needed.

A manual ball valve provides a means to isolate the entire monitoring station from the system water when needed.

Continuing downstream in the influent line is an assembly that serves to provide backflow prevention and then pressure regulation. The backflow-prevention device is a vented double-check valve. The pressure-regulating valve lowers the influent water system pressure to one specified for PRS Monitoring Station operation. The monitoring station is built to withstand and operate at 30 psig of pressure. The station pressure should be set to a level below the minimum influent system pressure. The goal is to keep a steady pressure on the monitoring station, which will, in turn, keep a steady flow rate through the test chambers. In cases where the minimum influent system pressure is below 30 psig, the monitoring station pressure can be set lower, but about 15 psig is the lowest pressure recommended to push the water through the metal plates of the test chambers.

An electrically actuated slow-closing valve is next in line. The valve is controlled by the timer that is mounted on the monitoring station frame. This is the means by which the monitoring station operates automatically and with a consistent flowing/ stagnating water schedule. If electricity is interrupted to the monitoring station, the valve will "fail closed" to prevent continued flow.

An inline pressure gauge displays the pressure of the monitoring station when the water is flowing.

The influent sample tap is downstream of the pressure gauge. Influent water that flows through the test chambers is sampled from this tap.

The influent sample tap serves a secondary purpose in preparing the monitoring station before the metal plates are installed. That is, a disinfecting solution is pumped into the monitoring station through a hose connected to the influent sample tap with the manual influent ball valve isolating the station from the water system. This is described in Section A.3.

A pipe coupling downstream from the influent sample tap allows for the removal of the influent line if cleaning or maintenance of the influent equipment is required.

A.1.2 BACKFLOW PREVENTION

Backflow prevention is an important issue to be addressed for monitoring station installations. Each state has its own regulations on backflow prevention requirements that should be consulted.

When the PRS Monitoring Station is used as a chemical testing station with chemicals dosed into the test chamber lines, there is the potential for back-siphonage of chemicals into the water distribution system. In Wisconsin, for example, this situation requires the use of a device with an ASSE 1056 certification. The device must be mounted upstream of, and above the height of, the monitoring station. Consult specific state regulations to determine if such a device must be registered, inspected, and a fee paid.

When the monitoring station is used for monitoring only, with no chemical feed, a vented double-check valve device (certified as ASSE 1012) is typically required and needs no registration. This valve has been described previously as a piece of equipment on the influent line.

A.1.3 TEST CHAMBER LINE

After water flows through the influent line, it is divided amongst the vertical lines that are associated with the test chambers. A check valve at the bottom of each vertical line prevents the water from flowing back into the influent line and mixing the water from the different test scenarios.

Next, a needle valve provides some sensitivity to tweaking and setting the flowrate of the water before it enters the test chamber. A totalizing flow meter, which is located downstream of each test chamber, provides the flow rate measurement used for setting the needle valve in the proper position. This will be discussed in the description of the monitoring station operations in Section A.5.

A tee with a ½-inch diameter female-threaded connection is installed downstream of the needle valve. The tee branch will only be used if a chemical is to be dosed into the test chamber line. Otherwise, it should stay plugged.

Downstream from the chemical injector tee is a static mixer. This is also used for chemical dosing scenarios to mix the chemical into the low flow entering the test chamber. The static mixer is not necessary for nondosing scenarios. It can be omitted in those cases and installed for later projects, if desired, if the proper vertical spacing is maintained.

Downstream from chemical injection and mixing is the test chamber. Below the test chamber in the vertical line is a union ball valve and above the test chamber is a pipe coupling in order to remove the test chamber completely from the vertical line if there is ever a need.

As mentioned previously, a flow meter is downstream of the test chamber before the water flows into the common discharge line.

A.1.4 TEST CHAMBER

A cutaway view of a test chamber is shown in Figure A.3. The test chamber is a 4-inch-diameter Schedule 80 PVC pipe that is 6.75 inches long. A flange is attached to each end.

As seen in Figure A.2, a blind flange is bolted to the bottom flange where water first enters the test chamber. The blind flange is drilled so that the vertical piping can be attached. In addition, a small hole is drilled for attachment of a short,

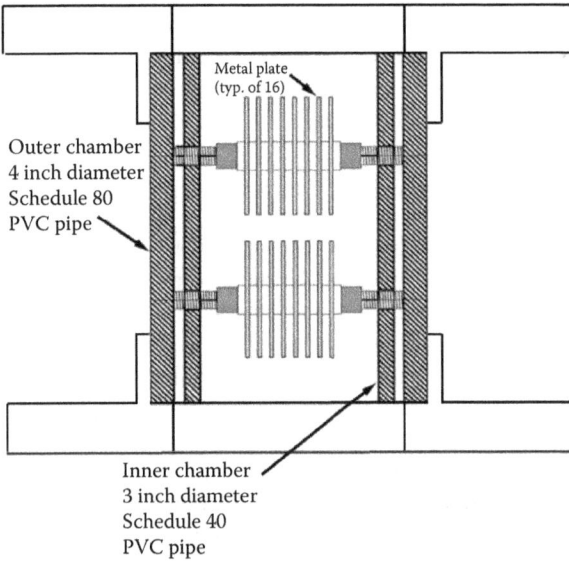

FIGURE A.3   PRS Monitoring Station test chamber cutaway view and metal plate holders.

small-diameter plastic pipe and manual ball valve. This serves as the sample tap for the test chamber and it draws from the bottom of the test chamber assembly.

A second blind flange is bolted to the top flange of the test chamber. A hole is drilled in the top blind flange so that the vertical piping can be attached and water can flow out of the test chamber past the flow meter and into the discharge line.

Inside the test chamber is a 3-inch-diameter Schedule 40 pipe cut into thirds. Matching notches are cut at the junction of each pipe section. There are two notches on each pipe section, located 180 degrees from each other. This configuration allows a plastic rod to be placed across the diameter of the pipe in the lower junction and another rod to be placed in the upper junction. See Figure A.3.

The plastic rods hold the metal plates that are important in recreating the metal surface area of a pipe in the test chamber. In order to obtain the proper surface area of metal-to-volume-of-water ratio, 16 metal plates of the dimensions 2 inches × 2 inches × 1/16 inch must be used. Eight such plates are held on one rod by means of a hole drilled in the middle of the plate. Plates are held apart from each other by means of Teflon® spacers. A plastic nut is threaded on each end of the metal plate "sandwich" to hold all the plates in place. Two rods of eight plates each are cradled in the internal pipe's notches.

The metal plates can be made of any metal or alloy that is of interest to study, especially materials that are representative of the distribution system being monitored. All plates in a test chamber must be of the same material.

The metal plates are not added to the test chambers until the PRS Monitoring Station has been installed, tested, and disinfected for operation. This is discussed later in Section A.3.

It is possible to modify the monitoring station's configuration as need arises, adding or changing equipment. It is recommended, however, that the test chamber configuration remain the same. In this way, a modified monitoring station would continue to be essentially the same as any PRS Monitoring Station; the design would remain standard and the data comparable.

A.1.5   DISCHARGE LINE

All test chambers flow into a common discharge line. There is an air vent just above the discharge line on the right side of the monitoring station, which must be open when sampling stagnating water from the test chambers. It is necessary to open the air vent with the manual ball valve in order to use atmospheric pressure to push out the water from the test chamber into the sampling tap at the bottom of the test chamber. An air filter is attached to the air vent in order to prevent microbiological contamination from entering the monitoring station from the air.

This air vent has a secondary purpose in the disinfection of the monitoring station in preparation for operation. It is discussed in Section A.5.

A second air vent is located on the left side of the discharge line. It stays closed unless a vacuum were to occur in the discharge line. A less-than-atmospheric pressure in the discharge line will open the air vent automatically to prevent back-siphonage from the drain line that is connected to the discharge line.

The end of the discharge line has a manual ball valve for isolating the monitoring station if needed. The end of the line is a ½-inch-diameter female-threaded pipe connection. A line to a wastewater drain can be attached either as a plastic pipe, tubing, or a hose with the associated adapter connected to the discharge line.

It is important to leave an air gap between the end of the drain line and the drain to prevent back-siphonage of wastewater into the monitoring station.

A.1.6   FRAME

A metal frame holds the piping and equipment of the monitoring station. The frames have been built to fit through standard-height and -width doorways.

There are locking casters on the frame to roll the monitoring station to the location where it will be installed. The casters should be in locked position at the installation location.

A.1.7   TIMER AND STATION AUTOMATION

Water flow and chemical feed for the PRS Monitoring Station is controlled automatically by a programmable timer. A schematic is shown in Figure A.4. The timer signals the electric actuator of the influent valve to open and close. An electrical outlet is also wired to the programmable timer. Electricity is available to the outlet when the influent valve is open and water is flowing; the electricity at the outlet is off when the influent valve has closed and water is stagnating in the station. This outlet is intended to control chemical feed pumps as it is desired to dose any chemicals when water is flowing and to stop dosing when water stagnates in the monitoring station.

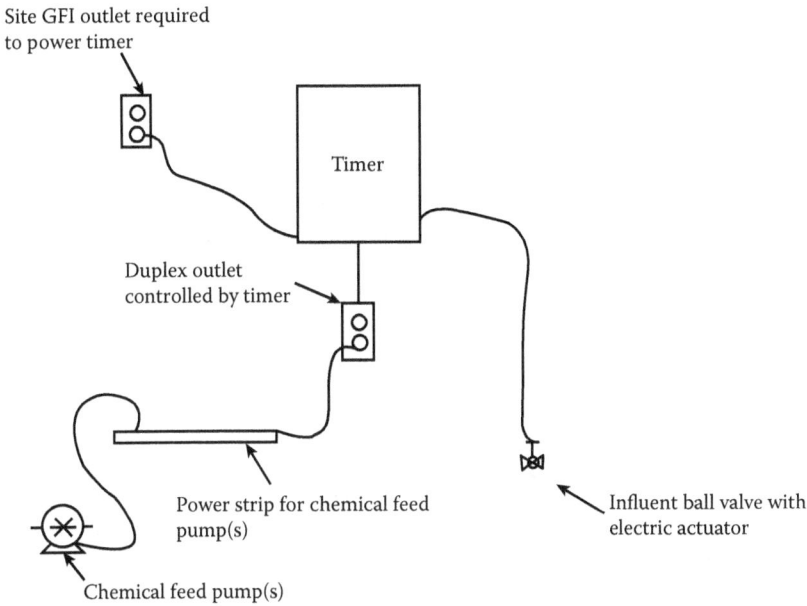

Site GFI outlet required
to power timer

Timer

Duplex outlet
controlled by timer

Power strip for chemical feed
pump(s)

Influent ball valve with
electric actuator

Chemical feed pump(s)

FIGURE A.4   PRS Monitoring Station timer and controls.

The main task in operating the PRS Monitoring Station is to keep a consistent flow rate through each test chamber. Each test chamber has its own flow meter with a flow-totalizing readout. The flow meters can be checked manually for the proper flow per day per test chamber. Or, flow rate and totalized flow can be downloaded electronically from each meter with an additional device to connect to a computer. The meters can also be wired to send the flow data over a SCADA system if one is available at the installation location.

One final possible automation is to use online sensors in a flowing water line sending the influent water to the PRS Monitoring Station. Typically, field tests (pH, temperature, oxidation/reduction potential (ORP), turbidity, conductivity, free chlorine, and total chlorine) are performed on the water influent to the monitoring station at least once a week. Online sensors can be installed on the line or a by-pass to the line providing the influent water and can send or store continuous data all week. Online sensors must still be calibrated, checked for accuracy, and cleaned weekly.

A.1.8 CHEMICAL FEED EQUIPMENT

Small package chemical feed systems work well for the PRS Monitoring Station. One such system is a 15-gallon solution tank with a peristaltic pump attached. The suction side of the pump is attached to tubing leading to a check valve attachment that can sit at the bottom of the solution tank, pull in the chemical solution, and prevent fluid from flowing back into the solution tank. On the discharge side, tubing connects the pump discharge to an injector that fits into the threaded injection tee on a vertical test chamber line. The injector also includes a check valve to prevent fluid from flowing back from the monitoring station into the chemical feed system.

Using a peristaltic pump, a dosage calibration system can be installed on the line between the pump discharge and the monitoring station injector. A plastic three-way bypass valve can be cut into the tubing so that the pump discharge can be diverted to a graduated cylinder to measure volume of fluid dosed over time.

A peristaltic pump is recommended over a positive displacement pump. A positive displacement pump requires the dosage calibration on the suction side of the pump, which has the potential for microbiological contamination of the PRS Monitoring Station.

A.2   STATION OPERATING PARAMETERS

The PRS Monitoring Station can be operated under many scenarios with a programmable timer to turn water flow on and off, just like water flows in building plumbing. However, certain operational criteria have been established for the PRS Monitoring Station in consideration of water conservation and in order to maintain standard operating conditions from project to project and water system to water system. A long stagnation time also exaggerates the chemical and microbiological interactions in the test chambers, making it easier to study water quality parameter changes.

To this end, PRS Monitoring Stations are operated at 0.5 gallons per minute of water flow through each test chamber; the flow occurs for only one hour a day. That is, 30 gallons (113.6 liters) of water are used per day for each test chamber in the PRS Monitoring Station. Pressure is kept constant in order to keep the flow steady. The pressure should be set below the minimum system pressure of the influent water but not more than 30 psig and not less than 15 psig.

A.3   INSTALLATION AND STARTUP

There are many detailed instructions for installing, starting up, and operating a PRS Monitoring Station. The general steps are simple and straightforward:

- Set the monitoring station in place.
- Connect the influent and discharge hoses/tubing/piping.
- Test the monitoring station for water flow and leaks at the operating pressure and during water stagnation.
- Disinfect the monitoring station.
- Prepare the metal plates and install with a brief disinfection period.
- Begin normal operation of the monitoring station.
- Sample water according to your project's monitoring plan.

When installation begins, keep the metal plates and test chamber internals protected and stored away until after transport, installation, flow and pressure testing, and disinfection of the monitoring station.

To install the monitoring station, transport it to the monitoring location. After setting the station at its final location, lock the wheels. Close the manual influent and effluent ball valves to keep the monitoring station isolated from the water system during installation.

Run a hose, tubing or hard plastic piping from a tap in the water distribution system piping to attach to the influent pipe connection of the monitoring station. An adapter will be needed for connecting to a hose or tubing. Run a hose, tubing, or piping from the final discharge pipe connection to a drain using an adapter at the connection, if required. Keep the end of the drain hose above the drain to provide an air gap for wastewater backflow prevention.

Plug the timer into an electrical outlet that has a ground fault interrupter. If the outlet does not have one, an accessory with a ground fault interrupter can be purchased to plug into the outlet. An electrical surge protector is also advisable to have at this outlet.

Open the manual influent and effluent ball valves and all valves that allow the water to flow through the station. Push the manual override button on the timer to open the electrically actuated ball valve and allow system water to enter the monitoring station.

Adjust the pressure to 30 psig or lower with the pressure-regulating valve. Keep the pressure below the minimum system pressure so that the monitoring station pressure will remain steady.

With water flowing through the station, check for leaks. If there are no leaks, shut off the water.

Program the timer for opening and closing the influent valve and watch that the automatic operation occurs properly. Set the timer to off.

On the day before startup, disinfect the monitoring station. To do this, fill a five-gallon bucket with a disinfecting solution of water and bleach (sodium hypochlorite) to a concentration of 10 to 50 mg/L free chlorine. A higher chlorine concentration could possibly damage some of the plastic equipment in the station. As an example solution, if using household bleach with 5.7% available chlorine, mix 5 mL of bleach into four gallons of water. The disinfection solution will be pumped using a utility pump into the influent sample tap with discharge from the air vent. To do this, remove the air filter from the air vent on the discharge piping and connect tubing to the hose barb on the air vent that reaches almost to the floor. Place the tubing into a second five-gallon empty bucket. Connect a hose to the suction side of the utility pump that runs from the disinfecting solution into the pump. Connect a hose to the discharge side of the utility pump that runs from the utility pump discharge to the influent sample tap. The hose connection will need to be cut from the end of the hose, a hose clamp placed on the end of the hose, the hose slipped onto the hose barb of the influent sample tap, and the hose clamp tightened. Close the manual influent and effluent ball valves. Pump the disinfecting solution into the monitoring station to fill. Excess solution will come out of the air vent tubing and into the bucket below the tubing. Turn off the utility pump and allow the disinfecting solution to sit in the monitoring station for 24 hours. Do not disconnect the disinfection solution setup.

The next day should be the day of the metal plate installation and the monitoring station startup. The metal plates must be cleaned of oxidized metal debris on their surfaces. To do this, the plates should be set into a 2% solution of an organic acid detergent such as Citranox®. Even though this is not a strong acid, take precautions with proper ventilation and protective clothing and eyewear, as if working with a strong acid. First, clean the internal pipe sections, rods, spacers, and nuts with the detergent and rinse thoroughly with distilled or deionized water. Set the internals nearby on a clean surface. Then, clean the metal plates, allowing them to sit in the detergent, rinsing them off with distilled or deionized water, and positioning them on the plastic rods. Use gloves and tweezers to handle the plates. Use the test chamber internals—the notched pipe sections—as cradles to hold and transport the cleaned plates back to the monitoring station (Figure A.5).

Back at the monitoring station, drain the test chambers of the disinfecting solution by holding a bucket under each test chamber sampling valve and opening the

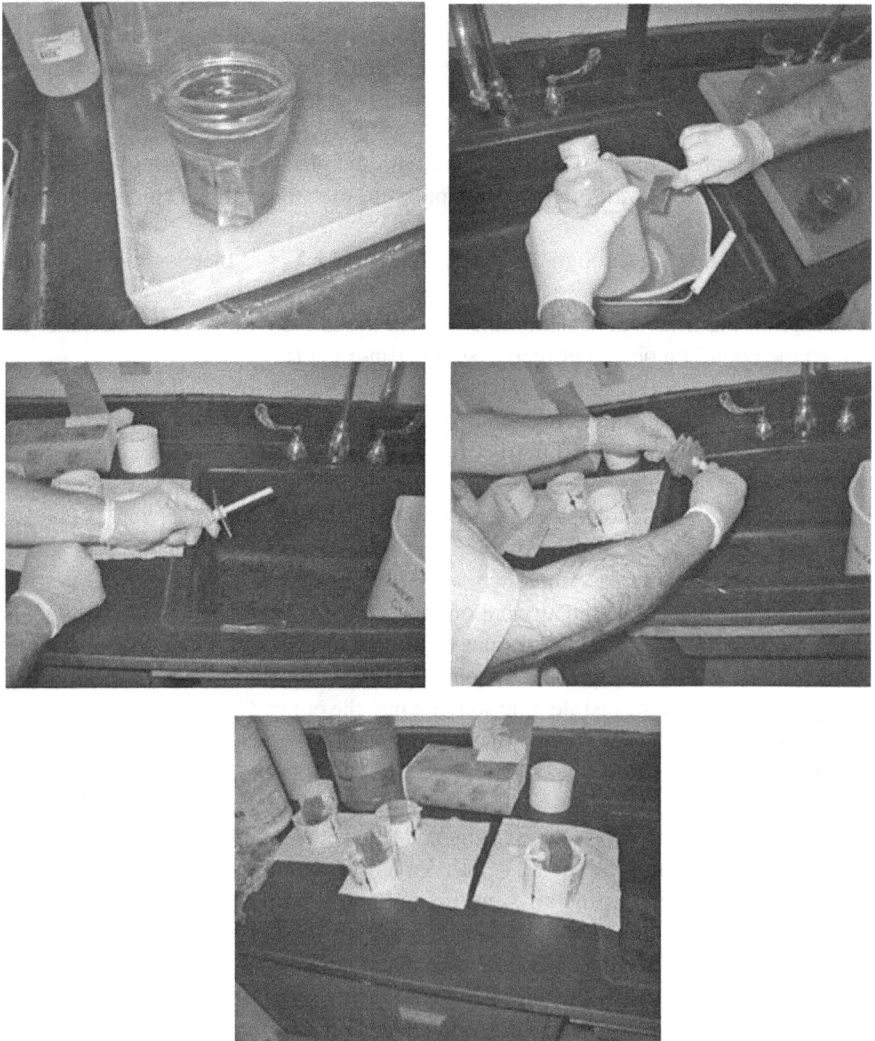

FIGURE A.5   Preparation of the plates for the PRS Monitoring Station.

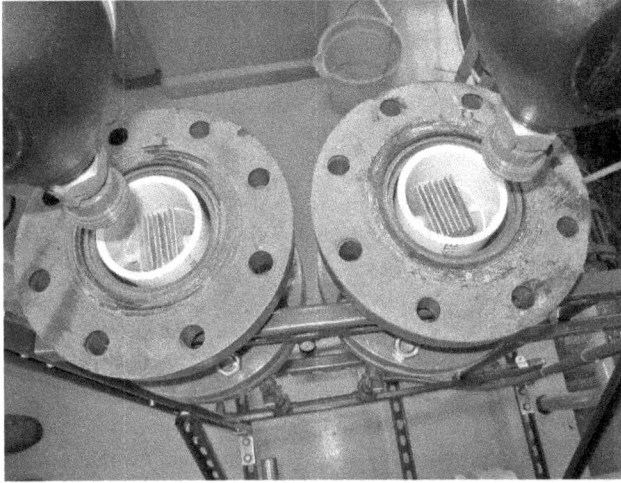

FIGURE A.6   Metal plates and internals set into test chambers of the PRS Monitoring Station.

sample tap ball valve. (Remember to close the sampling valve after the test chamber is drained.) Open the top flange of each test chamber. Remove the top flange section using the pipe coupling above the flange and possibly loosening the pipe clamp on the frame to allow some movement of the piping in order to remove the top flange section. Set the flange on a clean surface. Install the lower inner pipe section into the test chamber, set a rod of plates into the notches. Fit the middle section of the inner pipe over the rod, lining up the notches. Set the second rod of plates into the notches of the middle section. Place the remaining pipe section over the last rod (Figure A.6).

Reinstall the top flange and close up the test chambers. Pump in fresh disinfecting solution and allow it to sit in the monitoring station for 15 minutes. Disassemble the disinfection solution tubing setting valves back into forward system water flow position. Install the air filter. Override the timer and open the influent valve. Run system water to flush out the disinfecting solution. As the system water is running, recheck the monitoring station pressure. Set the test chamber flowrates by tweaking the needle valves and measuring flow per minute with the totalizing flow meters and a stopwatch to achieve 0.5 gpm per test chamber. Program the timer to allow water flow one hour a day. The hour selected for flow will depend on the sampling schedule (Section A.6).

A.4   CHEMICAL FEED SETUP

If chemical dosing is to be used in a test chamber line, set up the chemical feed package. Connect the chemical injector into the injector tee. Run tubing between the injector and the chemical pump discharge. Run suction tubing between the chemical pump suction side into a check valve that will sit at the end of the tubing in the chemical solution tank.

Prepare a diluted chemical solution that will produce a desired concentration of a chemical to be tested in a flow rate of 0.5 gpm of system water at a calculated pumping rate that the chemical feed pump is capable of achieving.

Gather information needed to calculate the dilution of the chemical product in the solution tank:

- The flow rate of the water through the test chamber. (For each test chamber line in the PRS Monitoring Station, the flow rate should be 0.5 gpm. This can also be stated as 30 gph.)
- The final dosage of the chemical of interest that is desired in the test chamber line water. (For this example, assume that 1 mg/L of phosphorus as P is desired in the water.)
- The maximum flow rate that the chemical feed pump is rated for and the midpoint of that range. (For this example, assume we have a chemical feed pump rated to deliver a maximum of 1 gph. The midpoint flow rate will be 0.5 gph.)
- The volume of the chemical solution storage tank. (For this example, assume a 10-gallon tank. This can also be stated as 37.85 liters.)
- The concentration of the chemical of interest in the chemical product to be tested. (For this example, assume that phosphoric acid is to be used where phosphorus is reported to be at a concentration of 160,000 mg/L as P.)

Now, the calculation can begin with some easy formulas for estimating the chemical product dilution. The formulas will not be explained, but they are based on "mass balances" around the test chamber line and around the chemical solution storage tank. The calculation is performed in two steps:

1. Estimate the final concentration of the chemical of interest in the chemical solution storage tank by multiplying the desired dose of the chemical by the test chamber line flow rate and dividing by the midpoint flow rate of the chemical feed pump. Make sure both flow rates have the same units. In our example, we will divide 1 mg/L P times 30 gph by 0.5 gph to calculate that our chemical solution storage tank should have 60 mg/L P.
2. Calculate the volume of chemical product to pour into the chemical solution storage tank by multiplying the final concentration of the chemical of interest in the chemical solution storage tank by the volume of solution to be held in the storage tank and divide by the concentration of the chemical of interest in the chemical product. In our example, we multiply 60 mg/L P, which we just calculated, by 37.85 liters and divide by 160,000 mg/L phosphorus in the chemical product. The answer is 0.014 liters or 14 milliliters (mL) of chemical product that should be added to the 10 gallons of water in the chemical solution storage tank. It is typical that less than a liter of the chemical product is needed. A graduated cylinder marked in milliliters is typically required to transfer the chemical product into the storage tank. The water used to dilute the chemical product can be water from the distribution system unless a chemical interference will occur. Alternatively, deionized metals-free water can be used.

Know the chemical product that you are working with. Have appropriate safety equipment readily available when working with chemicals. Acids and bases require an eyewash station nearby, protective gloves, safety glasses, and other protective clothing. Have proper ventilation and spill cleanup kits.

With the dilution calculation, the dosage is close to the desired amount, but probably needs some tweaking. Change the pump settings to achieve the desired dose.

Dosage verification should be performed often in chemical testing. The dosage will vary based on chemical feed pump settings, drifting away from the desired flow rate through the test chamber line, and on the age of the chemical solution. Dosage can be verified by sampling from water flowing through the test chamber from the test chamber sample tap. A second dosage verification is to test the dosing rate of the pump by directing dosed chemical solution to a graduated cylinder and measuring volume of solution delivered over a specific time period. This verification can be performed with a shutoff valve and a three-way sample valve cut into the discharge tubing from the dosing pump. (A peristaltic dosing pump is recommended because a positive displacement pump requires the calibration to be performed on the suction side of the pump and that becomes a possible source of microbiological contamination.)

Be aware of how long the solution in the chemical solution storage tank will last. For example, with a 10-gallon tank and a chemical feed pump set at 0.5 gph, one batch of solution will last 20 hours. The chemical feed pump runs for one hour a day, so this translates to 20 days before a new batch of chemical solution must be made.

A.5   OPERATIONS

The operation of the PRS Monitoring Station is straightforward. The timer schedule should be set to allow for one hour of flow per day. On days when stagnation sampling will take place, the sampling should take place six hours after the water flow has stopped. (Try to be consistent with the stagnation time before sampling.) Set the starting flow time so that the stagnation sampling time, seven hours later, will be convenient for operators. If a monitoring station is located where utility personnel are available all day, have the hour of flow occur in the morning of a workday so that flowing water can be sampled in the morning and stagnation sampling can occur in the afternoon of the same workday. If a monitoring station is located at a remote site, set the flow time seven hours before operators will be visiting the site.

The water flowing into the device is controlled by a timer that opens and closes a valve. It is important to verify that the proper flow of 0.5 gpm has occurred daily for one hour through each test chamber line. At least once a week, check the totalizer readings on the test chamber flow meters. Subtract the previous totalizer reading from the new one and divide by the number of days since the last reading to calculate the average gallons per day of water that has flowed through the monitoring station. The result should equal 30 gallons of water per test chamber line in the monitoring station. If the flow is incorrect, push the manual button on the timer to allow flow through the monitoring station. Record the beginning totalizer reading and the starting time. Record the totalizer reading again at the end of the time period. Calculate gallons per minute and achieve 0.5 gpm within a range of 0.45 to 0.55. Adjust the flow control valves on each test chamber line to adjust the test chamber flow at 0.5 gpm.

Turn off the monitoring station flow when set. Write down the new totalizer reading for the next time. (For convenience, keep a sheet of totalizer flow readings and dates near the monitoring station.)

At least once a week, verify the chemical dosages to each test chamber line if using chemical feed systems. Sample each test chamber sample tap during the set flow period. If necessary, the monitoring station can be turned on manually for flowing water and chemical feed for sampling. Always use a very low flow of water from the test chamber line sample taps to collect samples to obtain a more representative sample of the chemical dosage. In addition, a measurement of the chemical dosing rate can be performed by directing the chemical feed discharge to the bypass tubing and measuring the solution delivered to a graduated cylinder in a set time. This helps in adjusting the dosing rate of the pump.

Also check for leaks when visiting the monitoring station to prevent such a situation from becoming worse.

A.6   WATER SAMPLING

Follow the schedule of water sampling that was determined in the monitoring plan. Chapter 4 described the process of developing a monitoring plan.

Have any field analytical equipment and labelled field and laboratory sample bottles available to take to the sampling site. Appendix C describes typical field and laboratory analyses. Include any safety equipment for use with a field laboratory, such as eye wash stations, safety glasses, gloves, etc. Do not carry dangerous chemicals such as acids to a field laboratory unless absolutely necessary.

For water sampling at the PRS Monitoring Station, ideally, flowing water samples should be taken during the automatic flow period from the influent valve to the monitoring station. The flow will turn off automatically, as controlled by the timer. After six hours from when the flow has stopped, the operator should return to the monitoring station to take stagnation water samples from the test chambers.

Unfortunately, many water utility personnel cannot fit two visits to a monitoring station into their work day, so an alternative sampling protocol is performed. In the alternative protocol, the operator visits the monitoring station six hours after the automatic flow has stopped. Totalizer readings are written down for the calculation of the average flow per day since the last visit. Then, the test chamber sample taps are wiped with alcohol to disinfect them and stagnation water samples are drawn from the test chambers. After taking the stagnation samples, the flow is turned on manually and the pressure checked. If needed, based on the totalizer readings, flow rates are reset or tweaked and confirmed with a timed test on flow as recorded by the totalizers. Flowing water samples are taken based on the monitoring plan from the alcohol-wiped influent sample tap to the monitoring station. The timer will then be set back to automatic operation, turning off the flow. Final totalizer readings are written down to start a new week of automatic operation.

For flowing water samples, only allow a very low flow from the influent sample tap so as not to disturb the hydraulics of the monitoring station.

For stagnating water samples in the test chambers, keep a very low flow so as to not cause mixing of water from other parts of the monitoring station into the test chamber water. There is about 1 liter of water available in the test chamber for sampling. It is best to remove no more than 750 mL of stagnating water from the test chamber.

Immediately run any field tests on the samples that have been prescribed in the monitoring plan. Cap all sample bottles tightly. Label sample bottles. Store samples on ice in coolers. Fill out field sheets with times of sampling and analysis. Fill out laboratory sheets as requested by the laboratory, typically to identify sample locations, date and time of sampling, name of sampler, and water quality parameters to be performed on each sample. Note any field processing such as filtration or preservation with acid. Deliver or ship the samples to the laboratory as directed by laboratory personnel.

Back at the office, enter the field sheet data into the computer data management program as soon as possible. It is best to record everything—flow totalizer readings, sampling times, field test results, and comments on adjustments made in the field.

A.7   METAL PLATE FILM ANALYSIS

At the end of a planned monitoring period, the test chamber metal plates should be removed and sent to laboratories for chemical and microbiological analyses of surface accumulations, as described in Appendix B.

Special shipping containers are required to transport a rod of eight plates to each laboratory. The construction of a shipping container is shown in Figures A.7 and A.8.

There are two rods of metal plates per test chamber. One rod will be sent to a geology or materials laboratory; the other rod will be sent to a microbiological laboratory. For proper study of the accumulations, be careful not to touch metal plate surfaces.

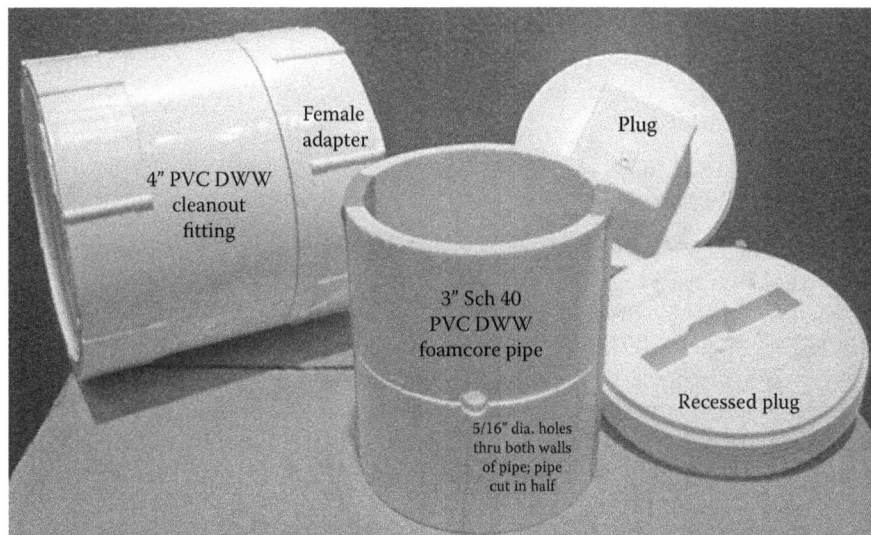

FIGURE A.7   Materials used for special metal plate laboratory shipping containers.

FIGURE A.8   Outside of a special metal plate laboratory shipping container.

For the rod that will be sent to the microbiology lab, pour about 15 mL of test chamber water into the shipping container to keep the metal plates moist. Pack the shipping container in a cooler with frozen gel packs and have the container delivered within two days to the laboratory.

For the rod that will be sent to the chemical analysis laboratory, there are no special requirements.

A.8   MAINTENANCE

At the end of a monitoring project, after a monitoring station is shut down and after the metal plates have been removed, the monitoring station should be cleaned, disinfected, drained, and dried.

Cleaning might be especially necessary if the water that the monitoring station was exposed to is prone to precipitating chemical scales, such as iron or manganese. A relatively gentle organic acid cleaning solution such as Citranox can be purchased from a laboratory supply catalog for this purpose.

Disinfecting of the monitoring station should be done using household bleach with no additives.

The method to deliver disinfecting solution to the PRS Monitoring Station was described in Section A.3. The same method can be used here for organic acid cleaning and for disinfecting.

The sections of the monitoring station can be disassembled for air drying. After drying, the station can be reassembled and stored away for the next use.

A.9   REUSE OF THE PRS MONITORING STATION

To start up another monitoring period with a previously used PRS Monitoring Station, purchase new plates, plastic holding rods, spacers, and nuts.

Start again at the installation and startup instructions.

Appendix B: Analyzing Pipe Wall Accumulations

Harvested distribution system piping and PRS Monitoring Station metal plates can be analyzed for surface accumulations both chemically and microbiologically for increased insight into the shaping of water quality in a specific water system.

The analysts of pipe films and scales can identify the existing compounds on surfaces that have been in contact with system water and can state how the compounds formed. That is, they can state what building blocks were present and what environmental conditions must have prevailed in the water system. That information describes the mechanisms most likely at work in shaping the distribution system water quality, including releasing lead and copper. From this information, hypotheses are strengthened as to what to manipulate in order to change the nature of the water system and the resultant water quality.

It is not only chemical reactions that affect water quality. Microorganisms play a highly important and complicated role. These tiny creatures can initiate or participate in chemical reactions. The nature of the biofilms that microorganisms form on surfaces must be defined.

The following analytical protocols are not typically performed commercially and are written for commercial and academic laboratories to adopt and begin providing as a service for water utilities. Chemical scales could be analyzed by commercial materials laboratories or university laboratories and other research laboratories, especially associated with geology and material science departments. Biofilms on surfaces could be analyzed by commercial water laboratories and special microbiological laboratories.

B.1 CHEMICAL SCALES

The protocols presented here were written by Dr. Barry Maynard, retired professor of geology at the University of Cincinnati. The protocols refer to the analysis of the PRS Monitoring Station metal plates but can be adapted to the analysis of harvested distribution system pipes, which need to be cut open lengthwise before analyzing.

B.1.1 STEP 1. INITIAL INSPECTION AND PHOTOGRAPHY

When first received in the laboratory, the shipper containing the plastic rod of PRS Monitoring Station metal plates is opened and the plates are dried in air. Since pipe films have developed on both sides of the metal plates, care must be taken to protect both sides when handling and storing the individual metal plates.

The metal plates are visually inspected. The distribution of scale textures and colors are described and recorded photographically. The scales will typically vary over the area of the metal plates, particularly at the outer edges and around the plastic spacers. These are areas of different flow velocity in the chamber and are likely to show differences in the crystal size of scale minerals.

Both macrophotographs, using a light stand, and photomicrographs, using a binocular microscope with up to 50× magnification, are needed. The metal plates are flat and good focus can be achieved across the whole field of view.

Because the human eye has a much better depth of focus than the camera lens, special care is needed to get good photographs, especially at higher magnifications. Use the strongest light source available. This permits the use of a minimum aperture on the camera lens, which improves depth resolution.

B.1.2   Step 2. Mineralogy by X-ray Diffraction

The most important scale characterization tool is x-ray diffraction (XRD) because it reveals the mineralogy of the dominant scale solids. Unlike pipes that have reacted for many years, the PRS Monitoring Station metal plates will generally not accumulate sufficient scale to be scraped off and analyzed by traditional powder diffraction methods. The analysis can be performed, however, by placing the metal plate directly in the diffractometer. The mineralogy determined can then be compared to the water chemistry to predict future reactions. Good diffraction records can be achieved with Cu K radiation with the conditions as shown in Table B.1.

The high-precision scan is conveniently run overnight. If machine time is limited, the rapid scan can be substituted, but the identification of minor phases and the estimate of relative mineral abundances are less reliable.

The flat plates are readily accommodated by the diffraction geometry of most instruments, but the sides of the plates may need to be cut down to fit the sample holder. This task is easily accomplished with a band or a scroll saw equipped with a metal cutting blade. However, both devices produce large amounts of metal shavings that need to be treated as hazardous material. Breathing protection and gloves should be worn. Cross-contamination between samples because of the cutting step happens easily. Careful cleaning of the saw between samples is essential.

Figure B.1 is an example of an XRD pattern from a lead metal plate. This sample is from an early stage of reaction and an unusually large variety of minerals are present, plus reflections from the underlying plate are very strong. A blank spectrum using an unreacted metal plate can be subtracted from this pattern to remove the lead metal peaks.

TABLE B.1
Diffraction Criteria for Metal Plates Using Cu K Radiation

Type of Scan	Time of Scan	Conditions of Cu K Radiation
High-precision scan	10 hours	5 to 75° 2θ, 0.01° step size, 5 sec. per step
Rapid scan	3 hours	10 to 60° 2θ, 0.02° step size, 4 sec. per step

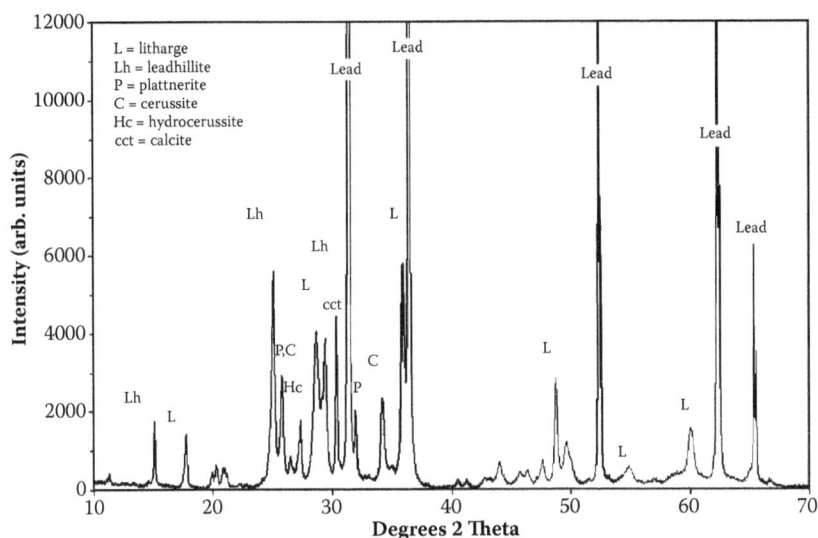

FIGURE B.1   X-ray diffraction pattern from a lead metal plate.

An estimate of relative abundance of the mineral phases can be obtained from the peak heights of the diffraction pattern. The largest peak, in the case shown in Figure B.1, the one for leadhillite at about 25 degrees, is assigned a value of 100. Then the peak height of the strongest peak for each of the other minerals is given as a percentage of the height of this peak.

B.1.3   STEP 3. MINERALOGY BY RAMAN SCATTERING

Raman spectroscopy is not employed as often as XRD in corrosion studies, but it is a valuable complementary technique because it can detect poorly crystalline phases and it provides confirmation of identifications made by XRD for minerals with closely overlapping patterns. The technique is widely used for characterization of art works and a variety of tools are available for widely varying prices.

One common system that is well-adapted to pipe scale analysis is a Renishaw inVia™ Reflex spectrometer with a 250-mm focal length system; similar Raman systems are available from other manufacturers. The Raman microscope exploits the Raman Effect to identify and characterize the chemistry and structure of materials in a noncontacting, nondestructive manner. The Raman Effect occurs when laser light is directed onto a material. Light is scattered, a tiny fraction of which is shifted in frequency as atoms in the material vibrate. Analysis of the frequency shifts (spectrum) of the light reveals the characteristic vibration frequencies of the atoms and hence the chemical composition and structure of the material. Particles as small as one micrometer can be uniquely identified.

Laser Raman is a technique that has not been widely applied to scale characterization, but shows great promise for distinguishing PbIV from PbII compounds. At an incident wavelength of 514 nm, strikingly different spectra are observed, as shown in Figures B.2 and B.3. Also, the PbIV pattern can be converted to the PbII pattern using high laser power irradiation, providing further confirmation of the presence of plattnerite (lead oxide where lead is in the +4 oxidation state) in the sample. That is, plattnerite has the interesting property that, under intense laser irradiation, it is quickly transformed to a lower- or even mixed-oxide mineral where the black color converts to light gray or light orange. At the same time, the characteristic PbIV Raman lines in the 500 wavenumber (cm^{-1}) region disappear, to be replaced by intensification of lines in the ~140 and 300 cm^{-1} regions, characteristic of PbII and mixed valance state lead oxides. For samples with large amounts of plattnerite, signal intensity can easily increase a million times after irradiation as compared to before intense laser irradiation. This transformation is irreversible and appears to be unique to plattnerite among all common lead

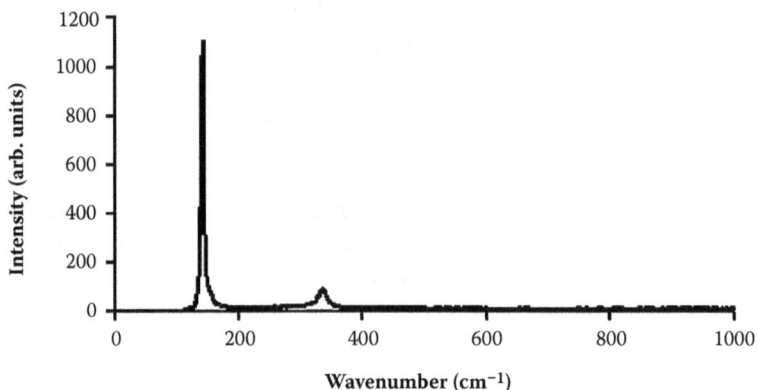

FIGURE B.2   Laser Raman spectrum for litharge (PbO) where lead is in +2 oxidation state.

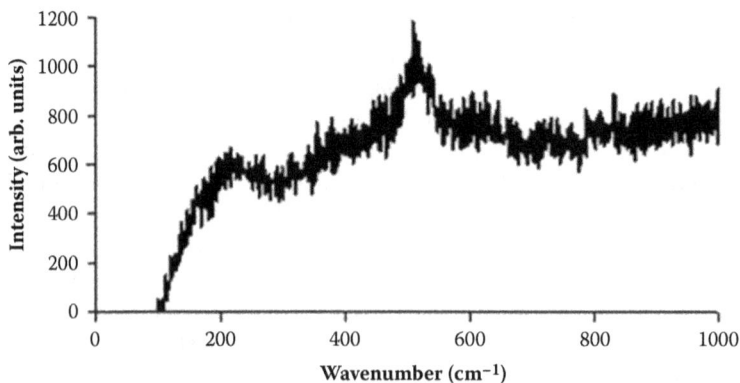

FIGURE B.3   Laser Raman spectrum for plattnerite (PbO$_2$) where lead is in +4 oxidation state.

minerals. Therefore, it provides unambiguous confirmation of the presence of plattnerite, even in small amounts.

B.1.4 Step 4a. Chemistry by X-ray Fluorescence

The flat nature of the metal plates makes x-ray fluorescence (XRF) a good choice for bulk chemistry determinations. The metal plate will need to be cut down to the size of the sample holder, as described earlier. Furthermore, the large area of metal will give exceedingly high blank values for lead or for copper, so only semiquantitative and comparative data are possible.

The blank issue can be reduced by running an unreacted metal plate and subtracting those values. Care must be taken, however, to ensure a clean, unoxidized surface for the metal plate blank. This is best achieved by washing the plate briefly in 10% nitric acid, followed by thorough rinsing with distilled water and then with acetone. A bright surface should result.

Useful elements to check for on the metal plates are: aluminum, barium, calcium, cobalt, chromium, copper, iron, magnesium, manganese, nickel, phosphorus, sulfur, silicon, uranium, and zinc.

Note that some barium peaks are severely overlapped by minor copper peaks. The strong copper radiation coming from a copper metal plate will give a very high barium blank for some analytical lines. Other elements may have lesser interferences, so it is essential to run a blank. Because much of the lead or copper measured will be from the metal plate rather than the scale, results cannot be reported as percentages of the scale by weight. It is more appropriate to report ratios of the counts for each element to the counts for lead or copper, after subtracting blank values for each element.

Note: Do not subtract the blank for lead or copper, which can produce negative values.

B.1.5 Step 4b. Chemistry by ICP or ICP Mass Spectrometry

An alternative approach (or a complementary approach if there is sufficient budget) is analysis by inductively coupled plasma (ICP) or by ICP followed by mass spectrometry, which is more sensitive. Some elements are not well covered by this technique, such as silicon and sulfur, but others such as arsenic are better represented. Again, the principal issue is working with a high blank from either lead or copper metal plates.

Rather than scraping the scale from the surface of the metal plate, it is recommended that the plate be immersed briefly in a fixed volume (typically 100 mL) of 10% nitric acid until the metal surface appears shiny. This solution can then be used directly in the ICP or diluted as appropriate.

B.1.6 Step 5. Microanalysis by SEM-EDS

Additional insight into scale chemistry can be obtained by using scanning electron microscopy (SEM) with energy dispersive spectroscopy (EDS). This approach extends to much higher magnification than light microscopy and can be used to study reaction sequences and, in some cases, to identify biological components of the scale. The energy dispersive analysis gives information about spatial distribution of chemical elements in the scale.

Sample preparation involves cutting a small section from the metal plate, about 1 cm square. Again, cross-contamination of samples occurs easily and cleaning the saw blade between samples is crucial. The samples are then mounted on an aluminum stub, coated with gold/palladium to prevent charge buildup, then placed in the instrument.

Typical EDS systems cannot distinguish certain similar compounds from each other. This is because quantification of very light elements such as carbon and oxygen is difficult. For example, cerussite and hydrocerussite, both lead carbonates commonly found in lead pipe films, cannot be differentiated by the EDS results. SEM is necessary to differentiate the compounds from each other by revealing their morphology. The lead oxides, litharge and plattnerite, are similarly difficult to distinguish. See Figure B.4 to see the similarity in the EDS patterns for lead oxides. Figure B.5 shows that litharge commonly occurs as hexagonal plates, whereas plattnerite tends to make prismatic crystals more tightly bunched. These minerals can also be differentiated with Raman spectroscopy through an optical microscope.

Images should be made in both the scanning and the backscatter mode. Scanning is a direct image, whereas backscatter intensity is a function of atomic weight of the atoms in a mineral. Backscatter intensity is proportional to Z, the average atomic weight of the elements present, and is very sensitive to small variations in composition. It is a valuable mapping tool for finding compositional differences. The needles in Figure B.6 are copper carbonate, whereas the fine, granular material with slightly higher backscatter intensity is copper oxide.

On occasion, biological structures are revealed in the SEM. Most bacteria are too small for effective imaging, but filamentous forms are distinctive, as are diatoms. See Figure B.7.

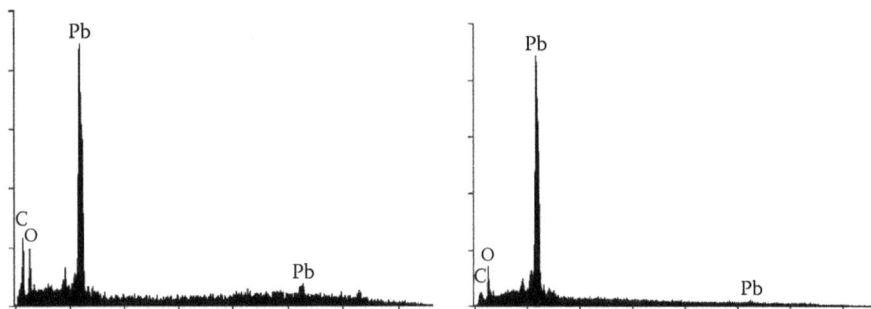

FIGURE B.4 Similar energy dispersive spectroscopy (EDS) patterns: litharge (PbO) (left) versus plattnerite (PbO_2) (right).

FIGURE B.5   Scanning images: litharge (PbO) (left) versus plattnerite (PbO_2) (right).

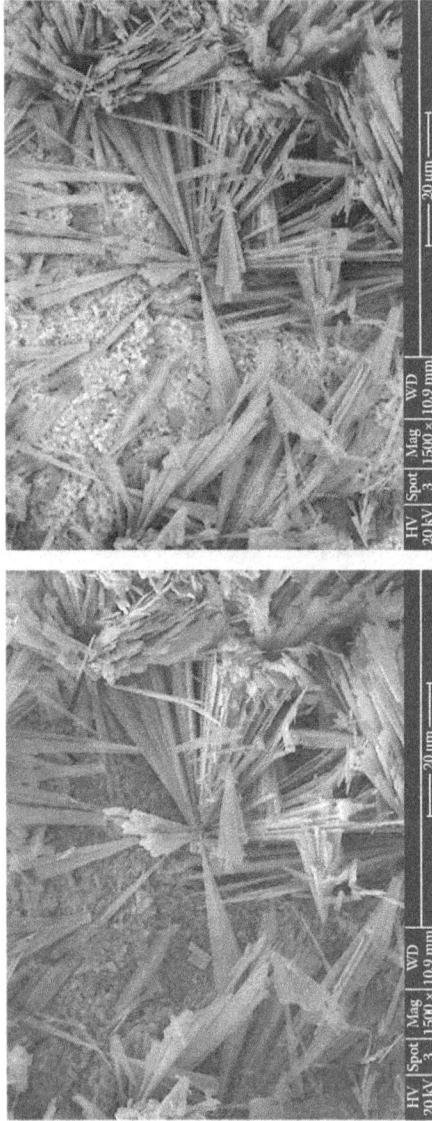

FIGURE B.6 Two ways of imaging in the SEM: scanning mode (left) and backscatter mode (right).

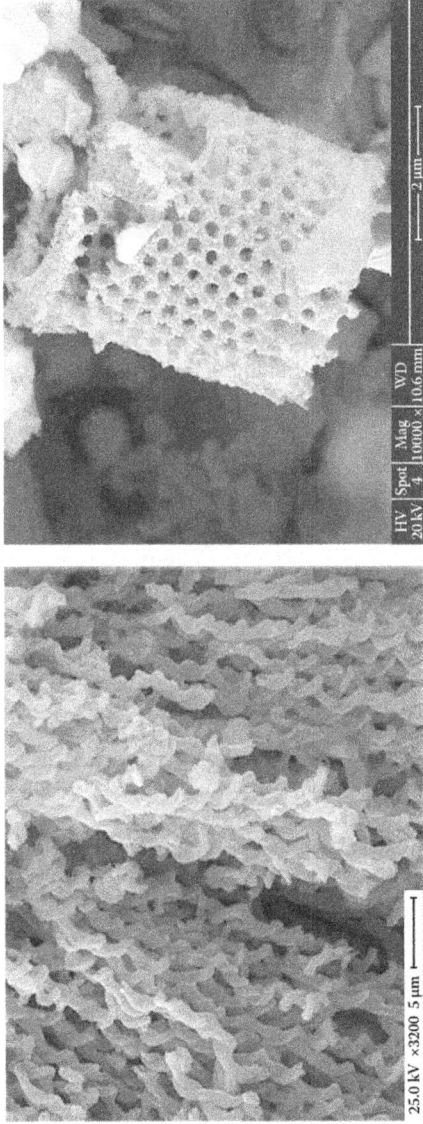

FIGURE B.7 Biological scale components: filamentous iron-oxidizing bacteria (left) and siliceous skeleton of a diatom (right).

B.1.7   STEP 6. REPORTING OF RESULTS

Results should be reported in the following five categories:

- Overall description—colors, textures (grain size, crystal shapes, etc.), extent of scale coverage. Include representative photographs with measurement scale shown.
- Bulk chemistry (XRF or ICP)—where the client needs information about the overall chemistry of the system. For example, are there additional elements with health implications such as arsenic or radium? Are there high concentrations of elements like manganese that are likely to become soluble if the oxidation state of the system is lowered by changing disinfectants? Results should be reported as a list of elements detected, ranked by amount relative to lead or to copper.
- Bulk mineralogy (XRD, Raman)—are the dominant scale minerals relatively insoluble, like plattnerite, or soluble, like litharge? Are there stabilizing minerals like calcite that serve to bind the scale more tightly? Conversely, a rapid buildup of calcite would be a preview of problems with scale buildup on heat exchangers and valves elsewhere in the system. The Raman data provide support for the XRD interpretations plus allow for the documentation of amorphous phases, which are common for aluminum, iron, and manganese. The results will be a list of minerals ranked by relative abundance.
- Spatial chemical changes (SEM-EDS)—is the scale evenly distributed on the metal plate? The issue for the client is whether the scale is forming a continuous protective film or is it spotty. As monitoring continues, the important question becomes, is there evidence of dissolution of earlier-formed scales? Results can be keyed to the photographs described in the first category of reporting above and should be a qualitative assessment of mineral distribution and reaction history.
- Spatial mineralogical changes (Raman, SEM)—provides information similar to the third reporting category, but includes insight into whether specific minerals are growing or dissolving or being overgrown by other minerals. The results will be in the form of a history of scale development in terms that can be compared to the treatment history of the system.

B.2   BIOFILMS

PRS Monitoring Station metal plates are studied microbiologically by means of submerging them in a lysing agent to release the biofilms. The cleaned metal plates are removed from the lysing agent and the final lysing solution analyzed for adenosine triphosphate (ATP), a measure of microbiological population. The result is reported in microbiological population per area of plate. This quantifies the degree to which biofilm has formed on the plates. Biofilm samples can also be taken from extra metal plates and sent for DNA analysis.

These protocols were developed by Dr. Andrew Jacque, President and Chief Scientist of Water Quality Investigations, LLC in Mount Horeb, Wisconsin. Dr. Jacque also includes several quality control steps in order to ensure that the biofilm has been completely removed from the plate surfaces. Harvested distribution system pipes can be analyzed in the same way but the pipes must be cut open while guarding against extraneous microbiological contamination.

It is especially informative in comparing the biofilm populations from plates of different metal types and different water systems. It is also informative in comparing the microbiological population in the water adjacent to the plates as a means to gauge the tendency of the microorganisms to stay on the plates in biofilms versus becoming entrained in the water.

Appendix C: Water Quality Measurement Guidelines

Water quality parameters and their sampling and analytical requirements are summarized in the National Primary Drinking Water Regulations (40 CFR Part 141 Section 141.23). Referenced in the regulation are analytical methods in publications of the EPA and several editions of Standard Methods for the Examination of Water and Wastewater (APHA et al. 1995). These references should be consulted for detailed information.

Laboratory personnel should also be consulted when planning a sampling event. In general, laboratory personnel determine the size of the water sample needed for each parameter. They also group certain parameters to be analyzed from the same water sample when there are common sampling and preservation requirements. The laboratory provides the sample bottles that have been prepared according to the analytical methods.

However, with the monitoring technique described in this book, some of the water quality parameters need to be considered in a special way. These requirements should be clarified with laboratory personnel before sample bottles are sent. Below is a list of common water quality parameter definitions and any special requirements related to this monitoring technique.

C.1 DESCRIPTION OF WATER QUALITY PARAMETERS

Adenosine Triphosphate
See "ATP."

Alkalinity, Total
Alkalinity is a composite of the concentrations of carbonate, hydroxide, and bicarbonate ions in the water. These ions are bases, so that, collectively, they represent the ability of the water to neutralize acids and resist pH change. Alkalinity is reported in units of mg calcium carbonate per liter (mg/L as $CaCO_3$).

Alkalinity can also be used to calculate the concentration of dissolved inorganic carbon (DIC) in the water. Special software is required to calculate DIC using alkalinity, pH, temperature, total dissolved solids (or conductivity), and hardness in order to determine the carbonate component of an alkalinity measurement. Based on DIC concentration, theoretical models of corrosion by-product formation and solubility have played a central role in advancing knowledge of uniform corrosion inside pipes. However, the models are not accurate or inclusive enough to be able to predict or quantify the corrosion of metals. Therefore, DIC has not been emphasized in this book.

Aluminum
See "Metals, Total and Dissolved."

ATP

A method to quantify microbiological populations is to measure the concentration of adenosine triphosphate (ATP) in the water. ATP is the energy molecule of living organisms. The measured concentration of ATP in the water is somewhat proportional to the number of microorganisms living in the water.

Only ATP from living microorganisms is captured in the analytical method by means of filtering the living microorganisms out of a water sample. Any ATP previously released from dead microorganisms is discarded in the water. The filtered living microorganisms become the sample to work with. They are exposed to a lysing agent that bursts the cells and releases the ATP into a liquid sample of the lysing agent. Another chemical compound is added to combine with ATP and emit light. The sample is placed in an instrument that can quantify the amount of light emitted and can correlate the measurement with ATP concentration.

Each type of microorganism has its own range of ATP concentrations per organism. As an estimate of microbiological population, an average ATP concentration per organism typically found in drinking water is used: 1000 microbial equivalents (ME) = 1 picogram (pg or trillionth gram) of ATP. While the actual type of microorganisms in the water is not known with this test, it is convenient to express the results as an estimated population number as it is useful for comparing and tracking the severity of microbiological growth in water systems.

The ATP test replaces the heterotrophic plate count (HPC) analysis, which has been used to indicate the general presence of microorganisms in water. However, HPC only measures the population of heterotrophic bacteria in the water. There are conditions where the HPC is low but other types of microorganisms are plentiful. This could only be detected with the ATP test that measures the presence of all microorganisms, except viruses, which do not carry their own ATP.

A standard of the EPA has been to consider less than 500 colony-forming units per mL of microorganisms as acceptable in drinking water. It is assumed, in that case, that there is enough disinfection to prevent excessive growth of microorganisms in the distribution system (40 CFR Part 141 Subpart H). This refers to results of the Heterotrophic Plate Count analysis, which, as stated previously, only identifies heterotrophic bacteria. Now that there is the ATP test that actually measures all microorganisms in the water, the standard of achieving less than 500 ME/mL (0.5 pg/mL) has been transferred to this new test by many practitioners. It becomes a more stringent criterion because all microorganisms are included in the ATP tests, not just the heterotrophic bacteria.

When taking a water sample for ATP, obtain a sterile sample bottle (typically, 100 mL) from a laboratory. Do not touch the bottle rim, the inside of the sterile cap, or the inside of the bottle. Fill the bottle to the indicated "fill line." Cap the bottle tightly and pack with frozen gel packs into a cooler. Deliver to the laboratory within 48 hours. Results are expressed in microbial equivalents per milliliter (ME/mL) or picograms of ATP per milliliter (pg/mL).

It is suggested that ATP tests be run on water that has stagnated in the water system, in addition to a flowing water sample, in order to observe how quickly microorganisms grow under stagnating conditions.

Calcium
See "Metals, Total and Dissolved." See also "Hardness, Total."

Carbon, Assimilable Organic
See "Carbon, Total Organic."

Carbon, Dissolved Inorganic
See "Alkalinity, Total."

Carbon, Dissolved Organic
See "Carbon, Total Organic."

Carbon, Total Organic
The total organic carbon (TOC) test rejects carbon combined in inorganic minerals, such as in calcium carbonate, and measures "all carbon atoms covalently bonded in organic molecules" (APHA et al. 1995). Organic carbon compounds are of concern in water systems because they can react with chlorine disinfection to form carcinogenic disinfection by-products. TOC is reported in mg/L as carbon (C).

A subset of total organic carbon is dissolved organic carbon (DOC). It is the carbon remaining in a water sample after filtration of a water sample through a 0.45-μm pore-size filter. The organic carbon remaining on the filter paper is in particulate form.

A subset of dissolved organic carbon is assimilable organic carbon (AOC). AOC is of concern because the carbon is in a form readily available to microorganisms as a nutrient for growth. Think about a human ordering lunch at a restaurant. Will that human order a whole cow or a hamburger? Answer: Hamburgers are easier to eat. AOC is the "hamburger" of the microbial world and is formed when organic carbon compounds are broken into smaller, more accessible molecules.

AOC is a very important parameter for tracking the biostability of a water system. The test involves a comparison of microbiological growth in the sample water to that in reagent water spiked with acetate, an organic carbon compound. Water samples must be delivered within 24 hours to the laboratory. AOC is reported in μg/L as acetate-C.

However, it is a very expensive test and only a few laboratories can perform the analysis. For this reason, the DOC test can be used instead. It is run routinely in most water laboratories. AOC can very roughly be estimated as one-tenth of the DOC. Given that the potential for excessive growth of microorganisms begins to increase over 50 μg/L of AOC (van der Kooij 1992), there is concern for excessive growth of microorganisms with a DOC greater than $50 \times 10 = 500$ μg/L DOC = 0.5 mg/L DOC.

Chloride
Chloride forms compounds of high solubility with metals. Therefore, its influence on the uniform corrosion of metals, the inhibition of which is dependent on the formation of low solubility metal compounds, must be considered.

There are relationships between chloride, sulfate, and alkalinity (carbonate concentration) in their competition for metal ions that are not discussed in this book.

Chlorine Dioxide

A few water systems use chlorine dioxide as a disinfectant. The appropriate field test kit must be used to measure for chlorine dioxide. That is, the proper reagents must be used and the colorimeter to read the resultant sample color must be calibrated for chlorine dioxide. Units are mg/L as ClO_2.

Chlorine, Free

See "Chlorine, Total."

Chlorine, Monochloramine

See "Chlorine, Total." See also "Nitrogen, Ammonia-."

Chlorine, Total

Total chlorine is the sum of free chlorine and chlorine combined in larger molecules. The free chlorine fraction is the sum of molecular chlorine, hypochlorous acid, and hypochlorite ions. It is the hypochlorous acid that is the most effective as disinfection in water systems. When water is analyzed for free chlorine, only the overall concentration is measured. Its effectiveness as a disinfectant is dependent on pH, where a lower pH produces more of the hypochlorous acid fraction (Connell 1996).

The combined chlorine fraction is typically composed of various nitrogen and chlorine compounds called chloramines. Chloramines include monochloramine, dichloramine, and nitrogen trichloride. To determine the concentration of combined chlorine, analyze for both total chlorine and free chlorine. Total chlorine minus free chlorine is the combined chlorine concentration.

For water systems that use free chlorine as a disinfectant, total chlorine results should equal free chlorine. If there is a difference between the two tests, this is an indication that organic and nitrogen compounds in the water system are combining and using up the free chlorine dosage.

For water systems that use monochloramine as a disinfectant, total chlorine results are typically representative of the monochloramine concentration.

There is yet a separate test for monochloramine that is of interest for water systems with chloramine disinfection. When using chloramine disinfection, the goal is to keep a large percentage of the combined chlorine in the form of monochloramine. Test results are confusing. Laboratory methods differentiate and report monochloramine as a chlorine species (mg/L as Cl_2). Field test kits sometimes differentiate monochloramine in the context of nitrogen compounds and report it as mg/L as NH_2Cl. When performing any arithmetic calculations between total chlorine and monochloramine, data must be in the same units. To avoid converting units, track the ratio of monochloramine to total chlorine, where any units can be used, just as long as the same units are always used.

The free chlorine and total chlorine tests have the units of mg/L as Cl_2. When reporting results, be specific as to which chlorine test was performed.

Chlorine tests must be performed on-site on water samples. The water sample can easily de-gas the chlorine when exposed to air. Take water samples in a glass container and fill the container to the brim. Cap the container immediately. Uncap only to perform the chlorine tests as soon as possible. Section C.2 explains the use of field kits to test for chlorine.

Conductivity

Conductivity is a measure of a water sample's ability to carry an electric current (APHA et al. 1995). Conductivity is higher in water with a higher concentration of inorganic compounds because there are more ion species available to carry a current. Therefore, conductivity is somewhat equivalent to the total dissolved solids (TDS) concentration. There is typically a general ratio between conductivity and TDS measured in a water system that can be determined with multiple measurements.

If conductivity or total dissolved solids concentration increases, this indicates that more compounds, especially compounds of metals, have dissolved in the water. It could indicate increased corrosion of metals or resolubilization of pipe wall metal compounds in changing conditions.

Conductivity is typically measured on-site using the proper electrode and meter. Results are reported in micro-Siemens per centimeter (μS/cm).

Copper

See "Metals, Total and Dissolved."

Disinfection by-products

Disinfection by-products (Total Trihalomethanes, Haloacetic Acids, Chlorite, and Bromate) are of concern in water systems and are regulated in the national primary drinking water regulations (40 CFR Part 141 Subpart L). Organic compounds can combine with chlorine intended for disinfection to create these carcinogenic compounds. Tracking the formation of these compounds can be incorporated into monitoring plans, as changes in water system chemistry for other purposes may affect their formation. Especially confirm compliance with disinfection by-product regulations when altering disinfection levels, pH, and total organic carbon concentration.

Hardness, Calcium

See "Hardness, Total."

Hardness, Total

Hardness is the sum of calcium and magnesium concentrations in the water. It is expressed in units of mg calcium carbonate per liter (mg/L as $CaCO_3$). The concentration of calcium in units of mg/L as Ca is converted to units of mg/L as $CaCO_3$ by using a ratio of molecular weights. One calcium carbonate molecule has the weight of one calcium atom (40), one carbon atom (12), and three oxygen atoms (3 × 16) and is equal to 100. The following ratio is used to convert a weight of calcium to a weight in terms of calcium carbonate:

$$\text{[the weight in terms of calcium carbonate]}/$$
$$\text{[the weight in terms of calcium]} = 100/40 = 2.5$$

Therefore, the weight in terms of calcium carbonate is 2.5 times the weight in terms of calcium only. For example, 94 mg/L as Ca is equivalent to 235 mg/L as $CaCO_3$. This result is also referred to as calcium hardness.

Magnesium concentration is converted in the same way where a magnesium atom has the weight of 24 and the ratio of calcium carbonate to magnesium is 100 to 24 or 4.2. For example, 45 mg/L as Mg is equivalent to 189 mg/L as $CaCO_3$.

Total hardness is the sum of calcium and magnesium hardness, and in this example, is equal to $235 + 189 = 424$ mg/L as $CaCO_3$.

Heterotrophic Plate Count

Heterotrophic plate count (HPC), a measure of heterotrophic bacteria in water, has been a useful analysis for decades in indicating the general presence of microorganisms. However, it is no longer necessary as an indicator because the ATP test measures the presence of all microorganisms, except viruses. Nevertheless, the HPC analysis will be discussed here as surface water regulations still require HPC data (40 CFR Part 141 Subpart H).

There are several methods to performing this analysis. In order to compare data from one sampling event to another, use the same method each time. If measuring HPC for compliance purposes, use the method required by the regulation.

For a more sensitive test, use the R2A growth media instead of the standard HPC media. In addition, it is suggested that HPC tests be run on water that has stagnated in the water system in addition to a flowing water sample in order to observe how quickly microorganisms grow under stagnating conditions.

For analysis, a water sample is serially, decimally diluted (1.0, 0.1, 0.01, 0.001, 0.0001, & 0.00001 mL), after which, volumes of each dilution are deposited into sterile Petri plates. For example, 1 mL is deposited directly from the sample container, then 0.1 mL is deposited onto a second set of Petri plates (0.1 dilution). For 0.01 dilution, add 1 mL of the original sample water to 99 milliliters of sterile buffer, mix thoroughly and deposit 1 mL of this dilution to a third set of Petri plates for a 1/100 (0.01) dilution. Deposit 0.1 mL of this same dilution to a fourth set of Petri plates for a 1/1000 (0.001) dilution, etc. All serial dilutions are analyzed in duplicate, and a control Petri plate containing only R2A agar is used to determine that the media is sterile.

After depositing serial dilutions onto Petri plates, add about 20 mL of R2A agar that has been prepared according to manufacturer's instructions and cooled to 44–46°C. The sample and agar should be gently mixed by carefully swirling or rocking the Petri plate.

The plates are allowed to cool, resulting in solidification at room temperature, and are then put inside a sealable container for five to seven days' storage at 25 ±3°C. The sealed container prevents excessive drying during incubation. For laboratories lacking a 25°C incubator, R2A plates may be stored at room temperature if the range is stable between 20–28°C (APHA et al. 1995). For laboratories lacking an area of reliable temperature but having a 35°C incubator, HPC agar may be substituted for R2A, and incubation time reduced to two days. However, the use of R2A agar is encouraged as the test is more sensitive to the actual presence of microorganisms.

When taking a water sample for HPC, obtain a sterile sample bottle (typically, 100 mL) from a laboratory. If the water to be sampled is chlorinated, make sure that a sodium thiosulfate "pill" is in the bottle to deplete the disinfection and protect microorganisms from its destructive effects. Do not touch the inside of the sterile cap or the bottle. Fill the bottle to the indicated "fill line." Cap the bottle tightly and pack on ice. Deliver to the laboratory within 24 hours. Results are expressed in colony-forming units per milliliter (CFU/mL).

Iron
See "Metals, Total and Dissolved."

Lead
See "Metals, Total and Dissolved."

Magnesium
See "Metals, Total and Dissolved." See also "Hardness, Total."

Manganese
See "Metals, Total and Dissolved."

Metals, Total and Dissolved
The typical water quality parameters mentioned in this book include a number of metals: aluminum, calcium, copper, iron, lead, magnesium, and manganese, for example. Metals can be found in the water in both dissolved and particulate (solid) form. The total concentration of a metal is the sum of the dissolved and particulate fractions. A water sample is acidified to measure for the "total" concentration of a metal. To analyze for the dissolved fraction of the metal, a portion of a water sample is first filtered through a 0.45-μm pore-size filter to remove metal particulates. The filtered sample is then acidified and analyzed and is reported as the "dissolved" concentration of the metal; the non-filtered portion of the sample is acidified and analyzed and is reported as the "total" concentration of the metal. To find the particulate concentration, subtract the dissolved concentration from the total concentration.

An important aspect that must be discussed with laboratory personnel is that the water sample should not come in contact with acid until a portion can be filtered for dissolved metal. The acid will quickly turn the particulate fraction into dissolved metals and the differentiation of metal forms will be lost. Make sure that the laboratory has sent bottles prepared for metal analysis but with assurance that any acid has been rinsed from them and the bottles are dry.

Sample bottle size must also be specific for this monitoring technique. When monitoring in residences, a one-liter wide-mouth sample bottle should be used, similar to the Lead and Copper Rule sampling protocol, to obtain the largest practical sample size that has been in contact with the residential plumbing during a stagnation period. The wide mouth of the bottle is necessary so that the flow of water from the faucet into the sample bottle can be similar to filling a glass of water; faster than that will atypically scour particulates off of the pipe walls; slower than that will not incorporate pipe wall debris that is typically captured in a glass of drinking water.

When monitoring the PRS Monitoring Station, a 250-mL sample bottle should be used on stagnating water in the test chambers and flowing water at the station's influent sample tap because of the limiting volume of the test chambers; sampling bottle filling flow rate is slower than at a faucet as conditions are different.

If looking for both total and dissolved metal concentrations, a portion of an individual water sample should be field-filtered. The unfiltered portion of the sample will be analyzed for total metal concentration and the filtered portion will be analyzed for dissolved metal concentration.

Filtration of the water sample should be performed on-site to best capture the actual concentrations of dissolved and particulate metal in the water sample. Very soon after sampling, the various fractions change from one to another as the water interacts with air. Particulate metal can also stick to the sides of the sample bottle. Field filtration protocols are discussed in Section C.2. If it is not practical to field-filter the water sample, deliver the water samples to a laboratory as soon as possible and ask for immediate laboratory filtration before acidification.

When looking for several different metals, the same water sample can be analyzed for all of the metals, typically, because they all require the same preparation and analytical method.

Some laboratories offer a long list of metals for one reasonable price per sample. In a number of PRS Monitoring Station projects, a quantitative inductively coupled plasma (ICP) scan has been run on each sample to measure 18 metals that are representative of typical piping system metals corrosion, treatment chemical additions, and source water metals, both in total and dissolved form. The metals are aluminum, arsenic, barium, cadmium, calcium, chromium, cobalt, copper, iron, lead, magnesium, manganese, nickel, potassium, sodium, strontium, tin, and zinc.

Because calcium and magnesium are of interest as background water quality to calculate the total hardness of the water, they are taken in flowing water at the influent to the PRS Monitoring Station. In most cases, there is only concern with the total concentration and there is no need for field filtration.

The limit of detection of the metals is important because metals at very low concentrations can be important to measure. Table C.1 lists preferable limits of detection.

TABLE C.1

Suggested Limits of Detection for Metals

Metal	Limit of Detection	Units
Aluminum	5	µg/L
Arsenic	0.5	µg/L
Barium	0.1	µg/L
Cadmium	0.1	µg/L
Calcium	0.15	mg/L
Chromium	0.5	µg/L
Cobalt	0.5	µg/L
Copper	1	µg/L
Iron	18	µg/L
Lead	0.1	µg/L
Magnesium	0.15	mg/L
Manganese	1	µg/L
Nickel	0.5	µg/L
Potassium	0.15	mg/L
Sodium	0.15	mg/L
Strontium	0.25	µg/L
Tin	0.1	µg/L
Zinc	5	µg/L

Many of the metals (such as lead, copper, iron, manganese) are best reported as micrograms per liter (μg/L). Other metals (such as calcium and magnesium) are typically found higher in concentration and are best reported as milligrams per liter (mg/L). Some laboratories report all metals in mg/L and the small numbers are difficult to work with. Convert mg/L to μg/L by multiplying by 1,000, where 1,000 micrograms equal a milligram. Be careful about converting back and forth between the two units and make the units clear on reports and calculations.

Nitrogen, Ammonia-
Nitrogen compounds are of interest in water systems because nitrogen is a nutrient that encourages the growth of microorganisms. One group of microorganisms uses nitrogen in ammonia (NH_3) to produce nitrite. A second group of microorganisms uses nitrogen in nitrite (NO_2^-) to produce nitrate (NO_3^-). This microbiological activity is called "nitrification."

It is important to know the concentrations of these chemicals in flowing water and how they change after stagnation. A measurement of ammonia-nitrogen concentration indicates the degree to which food is available for microbiological growth. In addition, measurements of nitrite-nitrogen and nitrate-nitrogen concentrations may indicate the degree to which nitrogen is being utilized by microorganisms. Nitrate can also enter the distribution system in the source water. All results are expressed as mg nitrogen per liter, but it is informative to indicate whether the test was for ammonia, nitrite, or nitrate as in mg/L as NH_3-N, mg/L as NO_2^-N, or mg/L as NO_3^-N. (Nitrite and nitrate can be reported as a combined concentration. This saves money on analyses but the trends of nitrite formation versus nitrate formation is lost.)

Tests for monochloramine, necessary to track the effectiveness of chloramine disinfection, is typically bundled with the ammonia and nitrogen field test kit. See "Chlorine, Total" for more details.

Nitrogen, Nitrate-
See "Nitrogen, Ammonia-."

Nitrogen, Nitrite-
See "Nitrogen, Ammonia-."

Oxidation-Reduction Potential
Oxidation-reduction potential (ORP) in water is the measure of water's capacity to undergo oxidation and reduction reactions. Corrosion chemistry is all about oxidation and reduction reactions; the solubility of metal compounds formed in the corrosion process are dependent on the ORP of the water. In addition, microorganisms have a difficult time growing and living under highly oxidative conditions. A high ORP indicates a water environment not conducive to excessive growth of microorganisms and it indicates the effectiveness of the disinfection present in the water.

ORP must be measured on-site using the proper electrode and meter. Results are reported in millivolts (mV).

pH
pH is roughly a measure of the concentration of hydrogen ions in the water. Hydrogen ions influence many chemical reactions, so pH is an important parameter to track. pH must be measured immediately in the field using an electrode and a meter. The field test is described in Section C.2. pH results are reported with the units of Standard Units (S.U.).

Phosphorus, Orthophosphate-
See "Phosphorus, Total."

Phosphorus, Total
Phosphorus can be combined in many forms in water. For the monitoring technique described in this book, the main focus is on total phosphorus and on phosphorus combined in orthophosphate ions. This is because phosphate-based corrosion control chemicals are used in the water industry. It is important to know if the phosphorus in the corrosion control products is in the form of orthophosphate or polyphosphate, a polymer chain of orthophosphate. Orthophosphate can combine with metal ions in water to form barriers on pipe walls that can inhibit further corrosion of metal pipes. Polyphosphates pick up metals in the water and as compounds on pipe walls and holds them in solution in the water. This increases the concentrations of metals, such as iron, manganese, lead, and copper, in the water. In addition, the properties of polyphosphates vary based on their molecular structure, which are typically proprietary. While we cannot know their structure or properties, we can at least know what percentage of our chemical product is a type of polyphosphate and what percentage is orthophosphate.

Another aspect is that polyphosphates eventually break up and revert to orthophosphate, changing the properties of the product that was purchased. It is informative to track the reversion of the polyphosphates to orthophosphates throughout the distribution system as lead, copper, and other metal concentrations in the water will be affected by this transition.

Therefore, have water samples analyzed for both total phosphorus and orthophosphate-phosphorus. The difference between total and ortho- is the polyphosphate concentration.

Be very careful about units of measurement. Total phosphorus is typically reported in mg/L as P (phosphorus). Orthophosphate-phosphorus is typically reported in mg/L as PO_4^{3-} (orthophosphate ion). To convert PO_4^{3-} to P, use the ratio of molecular weights as a conversion factor. That is, there is one phosphorus atom in an orthophosphate ion; phosphorus is about ⅓ the weight of the orthophosphate ion. 1 mg/L as PO_4^{3-} equals 0.33 mg/L as P.

Silicate, Silica, Silicon
Sodium silicate is sometimes used for corrosion control in water systems. The test for silicate concentration is reported as silica (SiO_2).

Solids, Total Dissolved
Total dissolved solids (TDS) in water, in simplified terms, represents the dissolved mineral compounds in the water. For example, for water with high alkalinity (carbonate anions) and hardness (calcium and magnesium cations), it would be expected

to see high TDS as well. Other minerals, such as iron and manganese, add to the TDS concentration. TDS is expressed in units of mg dissolved solids per liter or mg/L. See also "Conductivity."

Sulfate
As with chloride, sulfate forms compounds of high solubility with metals. Therefore, its influence on the uniform corrosion of metals, the inhibition of which is dependent on the formation of low solubility metal compounds, must be considered.

There are relationships between chloride, sulfate, and alkalinity (carbonate concentration) in their competition for metal ions that are not discussed in this book.

Temperature
Temperature is a factor in chemical reactions and in solubility of chemical compounds. Always record temperature during a sampling event as results may correlate to a significant temperature influence. Temperature must be measured immediately on-site when sampling. Use units of either degrees Fahrenheit (deg. F) or degrees Celsius (deg. C). Convert from one to the other using:

$$\text{deg. F} = (\text{deg. C} \times 9/5) + 32$$
$$\text{deg. C} = (\text{deg. F} - 32) \times 5/9$$

Turbidity
Turbidity represents the clarity of water caused by the presence or lack of suspended particulate matter. The matter can be from many origins, such as organic, inorganic, or microbial–reactive or inert. Portable turbidimeters have become a popular, inexpensive technique for tracking the cleanliness of the distribution system pipes and the quality of the distribution system water. When turbidity is on the increase, begin troubleshooting the system for incoming particulates to the water system or disturbance of existing pipe wall accumulations.

The units of measurement are in nephelometric turbidity units (NTU), which is based on the degree to which light is scattered by the particulate matter in the water. Turbidity must be analyzed immediately in the field with a portable turbidimeter or shortly after sampling with a bench-top turbidimeter.

Portable turbidimeters are also used in unidirectional flushing operations to determine an endpoint to the flushing, which is typically set at less than 1 NTU.

C.2   FIELD ANALYSES AND PROTOCOLS

There are some analyses that must be performed in the field just after obtaining a water sample because the water quality parameter changes very quickly with contact with air. Typical field analyses on drinking water are for measuring the water quality parameters: free and total chlorine concentrations, pH, temperature, conductivity, ORP, and turbidity. Convenient field test equipment is available for these tests.

All field equipment should be cleaned, calibrated, and checked at the office every day before going on site visits. Manufacturers' instructions should be followed. Below are some general comments on each test.

Chlorine Field Test Kits
Follow the manufacturer's instructions for maintenance and usage of the colorimeter for the analysis of various chlorine species. In general:

- Transfer 10 mL of sample water to a glass vial. Run this in the colorimeter as a blank and "zero" the instrument.
- Transfer another 10 mL from the same sample bottle to another glass vial.
- Add the appropriate powdered reagent for the analysis to the vial. Cap and shake the sample after adding the reagent. Use a stopwatch to wait for a prescribed and consistent reaction time before taking a reading on the colorimeter.

Be careful about whether you are measuring free chlorine or total chlorine. Free chlorine uses a different reagent than total chlorine. Also, the total chlorine sample with the reagent added must sit longer than the free chlorine sample before taking the reading.

For chlorine dioxide, the colorimeter must be calibrated to measure it and the proper reagent must be used. Consult with the manufacturer's instructions for proper procedures.

Field test kits for measuring monochloramine are typically included with nitrogen and ammonia field tests. The technique also uses a colorimeter after reagent addition. Follow the equipment manual for performance of this test.

Conductivity
Follow the manufacturer's instructions for maintenance and usage of the conductivity probe and meter. Typically, the probe must be stored in a special solution to keep the parts wetted and ready for measurement.

Calibrate the conductivity meter routinely depending on the frequency of use. When using the meter once a week, calibrate weekly before sampling.

In using the conductivity meter, rinse the probe with distilled or deionized water and pat dry with clean, lintless paper before putting the probe in the water sample. The meter may take time to stabilize in the range of the sample before displaying the results. Taking three measurements and using the average of the last two is a good practice.

pH and Temperature
Follow the manufacturer's instructions for maintenance and usage of the pH and temperature probe and meter. Typically, the pH probe must be stored in a special solution to keep the parts wetted and ready for measurement.

Calibrate pH meter routinely depending on the frequency of use. When using the meter once a week, calibrate weekly before sampling. To calibrate the instrument, use three standards; pH 4, 7, and 10. Change the standard solutions routinely. Especially change the pH standard solutions if the calibration line slope, as stated by the meter manufacturer, is not in the correct range.

In using the pH meter, rinse the probe with distilled or deionized water and pat dry with clean, lintless paper before putting the probe in the water sample. The meter may take time to stabilize in the range of the sample before displaying the results. Taking three measurements and using the average of the last two is a good practice.

Very low-alkalinity water will need a special pH probe that is sensitive to water where pH can easily fluctuate. Consult the equipment manufacturer for the proper probe.

Field Filtration

Filtration is performed on a portion of a water sample intended for metal analysis. The unfiltered portion is used for total metal analysis; the filtered portion is used for dissolved metal analysis.

Filtration should be performed as soon as possible after sampling to prevent on-going chemical reactions from altering the dissolved and particulate metal fractions.

Two sample bottles for metal analyses should be obtained from the laboratory for each intended water sample. There should be no acid in either bottle as the acid will destroy the ability to separate particulate metals from dissolved metals.

There are two options for field filtration. One method uses disposable equipment. The other method uses standard laboratory filtration equipment, which must be rinsed with nitric acid and metals-free deionized water in order to prevent cross-contamination between water samples.

If field sampling cannot be performed, make arrangements with the receiving laboratory to filter a portion of each sample as soon as possible.

A simple field filtration technique using disposable equipment is to use a plastic syringe with attached nylon filter. The syringes are equivalent to Norm-Ject® (Henke Sass Wolf, Tuttlinger, Germany), 50 mL and sterile with a luer fitting. The syringes are distributed by Air-Tite Products Co. (www.air-tite.com). The filters are equivalent to those manufactured by Fisher Scientific (Fisherbrand 25 mm Syringe Filter, 0.45 micron, nylon, non-sterile, Cat. No. 09-719F). A new filter and syringe must be used for each water sample to prevent cross contamination of metals between samples.

It is recommended that a burette stand with test tube clamp be used to grasp the syringe while pushing the water sample through the filter, but this is optional. Always hold the filter securely onto the syringe with one hand while pushing the plunger with the other or make sure that the filter is locked into place.

Always push about 10 or 20 mL of sample through a new filter and let the filtrate run to a waste bucket. This is to wet the filter with the sample so that there is no short-circuiting of sample around the filter during the actual filtration.

The filtration protocol is as follows:

- Attach an unused syringe and filter to a test tube clamp on a burette stand (optional).
- Shake the capped sample bottle holding the total metals sample.
- Remove the filter and plunger from the syringe. Reattach the filter and hold it on the syringe with one hand, or lock it in place if possible. Pour about 50 mL of sample into the syringe from the main water sample bottle, cap the bottle with one hand and set the bottle in a safe place.
- Put a waste bucket under the filter and plunge 10 to 20 mL of sample through the filter.

- Move the waste bucket out of the way and place an unused water sample bottle under the filter to filter the rest of the sample in the syringe.
- Refill the syringe as needed to filter about 100 mL of sample. This will take coordination, keeping the filtered water sample bottle in safe locations while the syringe is refilled, the filter is secured, and the plunger is set in place. Practice this before the real sampling event.
- After filtering, make sure that both the unfiltered water sample bottle and the filtered water sample bottle are securely capped.
- Properly dispose of the used syringe and filter.
- Mark the sample bottles with the proper sample identification indicating if the sample is field-filtered. The unfiltered water sample should be marked for total metal analysis; the filtered water sample should be marked for dissolved metal analysis.
- Store samples on ice and send to the laboratory where nitric acid will be added to preserve them.

For a reusable equipment method, standard laboratory filtration equipment is used: filtration flask, filter holder, filters, vacuum pump, vacuum tubing.

A 6% to 10% nitric acid solution must be used to rinse the filtration flask and filter holder. Acid must be used with caution and steps must be taken to protect human health as well as surrounding surfaces. In working with acid, take safety precautions as described below:

- Work in a well-ventilated area.
- Use gloves that are resistant to nitric acid.
- Use eye protection.
- Wear long sleeves, long pants, shoes, and an apron to protect skin.
- Have an eye wash available.
- Have running water or stored water available to wash acid from skin.
- Have baking soda available to neutralize spills.
- Capture acidic wastewater in a bucket partially filled with water and baking soda to neutralize waste before pouring it down a drain. pH test strips or a pH meter and probe can be used to indicate when wastewater is properly neutralized.
- Protect surrounding surfaces from spills.
- Use acid-resistant containers.

For each sample to be filtered:

- In a laboratory or work room, rinse filtration flask and filter holder with a 6 to 10 percent nitric acid solution, followed by several rinses with metals-free deionized water so that the equipment is at a neutral pH.
- When equipment is dry, seal all openings with a waxed laboratory film to keep the equipment clean during transport. Try not to acid-rinse at sampling sites because of the danger of ruining countertops and other surfaces.

If field rinsing is necessary because more than one sample will be filtered, set up a rinsing station complete with portable eye wash and gallons of emergency dilution water.

In the field:

- Gently shake the original capped sample bottle.
- Pre-rinse a new 0.45-μm nylon filter by filtering 50 mL of sample water through it.

TABLE C.2

Example Calculation of Field Analysis Accuracy Using Percent Recovery of Standard Solutions Measurement

Date	Standard Value	Measured Value	% Recovery	Moving Range
10/6/2014	1000	999	99.9	
10/13/2014	1000	1001	100.1	0.2
10/20/2014	1000	998	99.8	0.3
10/27/2014	1000	999	99.9	0.1
11/3/2014	1000	1000	100.0	0.1
11/17/2014	1000	1000	100.0	0.0
11/24/2014	1000	1001	100.1	0.1
12/1/2014	1000	1005	100.5	0.4
12/8/2014	1000	1000	100.0	0.5
12/15/2014	1000	998	99.8	0.2
12/29/2014	1000	1000	100.0	0.2
1/5/2015	1000	996	99.6	0.4
1/12/2015	1000	999	99.9	0.3
1/12/2015	1000	1002	100.2	0.3
1/26/2015	1000	1001	100.1	0.1
2/2/2015	1000	999	99.9	0.2
2/9/2015	1000	997	99.7	0.2
2/16/2015	1000	1002	100.2	0.5
2/26/2015	1000	1001	100.1	0.1
3/2/2015	1000	1003	100.3	0.2
		Average	100.0	0.233
		Sigma Unit	= Average Moving Range/1.128 = 0.207	
		Upper Control Limit	= Average (%) Recovery + 3 × Sigma Unit = 100.62%	
		Lower Control Limit	= Average (%) Recovery − 3 × Sigma Unit = 99.38%	
		Accuracy of Conductivity Analysis	100.0% accurate on average expected 99% of the time between 99.38% to 100.62%	

- Dispose of rinse water in a waste container and shake excess water from the filtration apparatus.
- Filter a portion of the sample.
- Pour filtrate into an appropriately marked clean sample bottle.
- Dispose of used filter.

Oxidation-Reduction Potential (ORP)

Follow the manufacturer's instructions for maintenance and usage of the ORP probe and meter. Typically, the probe must be stored in a special solution to keep the parts wetted and ready for measurement.

Calibrate the ORP meter routinely depending on the frequency of use. When using the meter once a week, calibrate weekly before sampling.

In using the ORP meter, rinse the probe with distilled or deionized water and pat dry with clean, lintless paper before putting the probe in the water sample. The meter may take time to stabilize in the range of the sample before displaying the results. Taking three measurements and using the average of the last two is a good practice.

Turbidity

Turbidity is a measure of particulates in the water. In a turbidimeter, light shines through the water sample and the deflection of the light by particulates in the water is measured. Any imperfections in the sample vial used in the turbidimeter can also scatter light and erroneously increase the measurement. This is the reason that it is

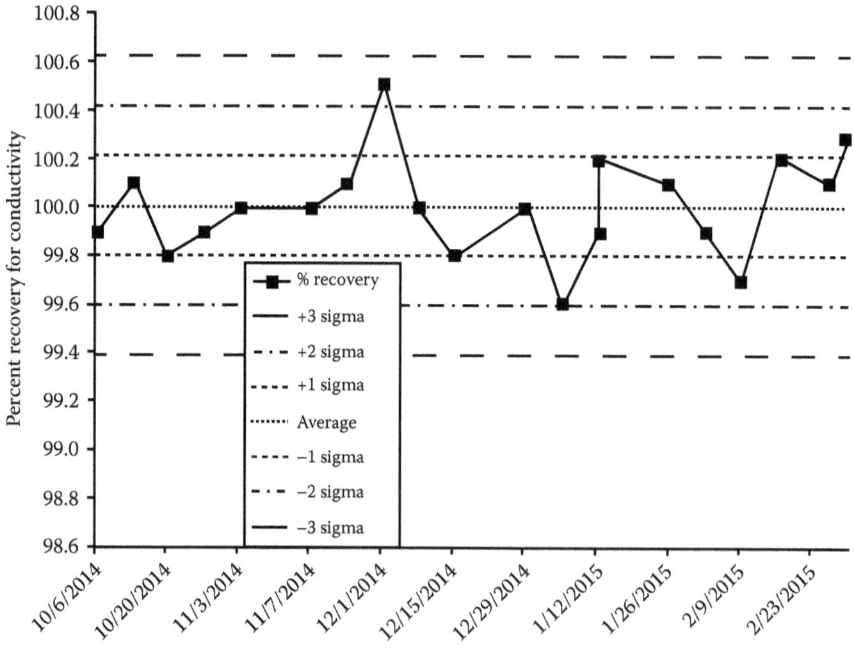

FIGURE C.1 Graph of Table C.2 Accuracy Calculations.

recommended that turbidity vials be coated with silicone oil before each measurement and polished smooth with a lintless cloth.

Turbidimeters must be calibrated according to the manufacturers' literature. In drinking water, there is a need to measure turbidity as low as 0.1 NTU up to 1.0 NTU and to also be able to accurately measure turbidities greater than 1 NTU, typically from 1 to 10.

C.3   QUALITY CONTROL OF FIELD ANALYSES

Laboratories with certification for drinking water analysis perform quality control on analyses for accuracy and precision (APHA et al. 1995). Although not typically done, quality control should be performed on field tests.

TABLE C.3
Example Calculation of Field Analysis Precision Using Duplicate Measurements

Date	Initial Measurement	Duplicate Measurement	Range
1/5/2016	515	522	7.0
1/12/2016	514	509	5.0
1/19/2016	510	512	2.0
1/26/2016	368	371	3.0
2/2/2016	512	512	0.0
2/9/2016	522	523	1.0
2/16/2016	533	525	8.0
2/23/2016	510	513	3.0
3/1/2016	519	523	4.0
3/8/2016	510	505	5.0
3/15/2016	514	514	0.0
3/22/2016	523	524	1.0
3/29/2016	520	520	0.0
4/5/2016	525	526	1.0
4/12/2016	520	518	2.0
4/19/2016	516	525	9.0
4/26/2016	519	522	3.0
5/3/2016	525	523	2.0
5/10/2016	511	513	2.0
5/17/2016	517	515	2.0
		Average Range	**3.0**
			= 3.286 ×
	Upper Control Limit on Range Chart		Average Range
			= 9.858
	Precision of Conductivity Measurement		+/− 9.86 μS/cm

For accuracy, a standard solution, where possible, should be measured for every 10 to 20 samples analyzed to determine how close the measurement can come to the known value. Percent recovery of the standard is calculated as:

Percent Recovery = 100 * (Measured Value)/(Known Standard Solution Value)

Percent recovery is then graphed as a Shewhart Control Chart (Wheeler and Chambers 1992). The average percent recovery is used as the accuracy of the analysis with the upper control limit and lower control limit describing the accuracy range that was achieved. A calculation and graph for accuracy using percent recovery of a standard is shown in Table C.2 and Figure C.1. This is a similar calculation as demonstrated in Table 5.3 and Figure 5.7, where subgroup size n=1.

Precision of an analysis is calculated by performing two measurements on the same sample for every ten to twenty samples measured. The absolute values of the differences between the first and second measurements were graphed as a Shewhart Range Control Chart (Wheeler and Chambers 1992) and the upper control limit used as the precision of the analysis. A calculation and graph for precision using duplicate measurements is shown in Table C.3 and Figure C.2. Here, the subgroup size n=2 and the graph is a plot of the range (difference between the two measurements) over time.

These quality control statistics and graphs call attention to any issues of equipment failure or improper technique so that they can be remedied quickly.

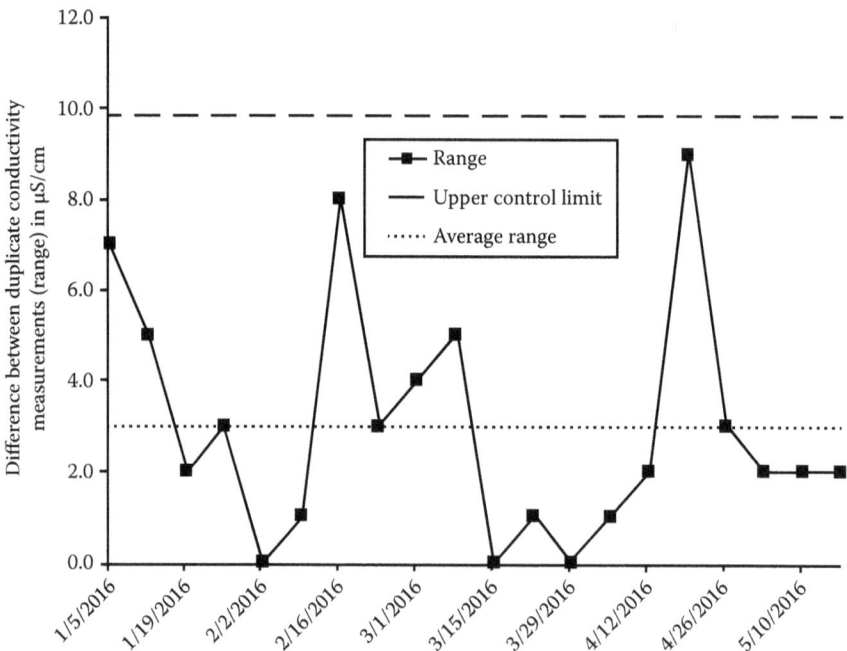

FIGURE C.2   Graph of Table C.3 Precision Calculations.

References

American Public Health Association (APHA), American Water Works Association (AWWA), and Water Environment Federation (WEF). 1995. *Standard Methods for the Examination of Water and Wastewater.* 19th ed. Edited by Andrew D. Eaton, Lenore S. Clesceri, and Arnold E. Greenberg. Washington, DC: American Public Health Association.

American Water Works Association (AWWA). 2017. *Manual of Water Supply Practices— M58: Assessment and Control of Internal Corrosion and Metals Release in Drinking Water Distribution Systems.* 2nd ed. Denver: American Water Works Association.

American Water Works Association Research Foundation (AwwaRF) and DVGW-Technologie-zentrum Wasser (DVGW-TZW). 1996. *Internal Corrosion of Water Distribution Systems.* 2nd ed. Denver: American Water Works Association Research Foundation/ American Water Works Association.

Bremer, P.J., B.J. Webster, and D.B. Wells. 2001. Biocorrosion of copper in potable water. *J. AWWA* 93:8:82.

Burlingame, G.A., D.A. Lytle, and V.L. Snoeyink. 2006. Why red water? Understand iron release in distribution systems. *Opflow* 32:12:12.

Cantor, A.F. 2006. Diagnosing corrosion problems through differentiation of metal fractions. *J. AWWA* 98:1:117.

Cantor, A.F. 2008. Pipe Loop on Steroids! A New Approach to Monitoring Distribution System Water Quality. *Proceedings of the American Water Works Association 2008 Water Quality and Technology Conference.* Denver: AWWA.

Cantor, A.F. 2009. *Water Distribution System Monitoring: A Practical Approach for Evaluating Drinking Water Quality.* 1st ed. Boca Raton: CRC Press.

Cantor, A.F. 2017. *Project 4586: Optimization of Phosphorus-Based Corrosion Control Chemicals Using a Comprehensive Perspective of Water Quality.* Denver: Water Research Foundation.

Cantor, A.F., J.B. Bushman, and M.S. Glodoski. 2003. A new awareness of copper pipe failures in water distribution systems. *WQTC Proceedings* (Philadelphia, PA; Nov. 2003). Denver: American Water Works Association.

Cantor, A.F., J.B. Bushman, M.S. Glodoski, E. Kiefer, R. Bersch, and H. Wallenkamp. 2006. Copper pipe failure by microbiologically influenced corrosion. *Mater. Perform.* 45:6:38.

Cantor, A.F. and J. Cantor. 2009. In Dr. Deming's Footsteps: Distribution System Water Quality Control and Process Improvement. *Proceedings of the American Water Works Association 2009 Water Quality and Technology Conference.* Denver: AWWA.

Cantor, A.F., D. Denig-Chakroff, R.R. Vela, M.G. Oleinik, and D.L. Lynch. 2000. Use of polyphosphate in corrosion control. *J. AWWA* 92:2:95.

Cantor, A.F., E. Kiefer, K. Little, A. Jacque, A. Degnan, J.B. Maynard, D. Mast, J. Cantor. 2012. *Distribution System Water Quality Control Demonstration.* Denver: Water Research Foundation.

Cantor, A.F., J.K. Park, and P. Vaiyavatjamai. 2003. Effect of chlorine on corrosion in drinking water systems. *J. AWWA* 95:5:112.

Code of Federal Regulations (CFR). Chapter 40 Protection of the Environment; Part 141— National Primary Drinking Water Regulations; Section 141.23—Inorganic chemical sampling and analytical requirements. Washington, D.C.: U.S. Government Printing Office.

Code of Federal Regulations (CFR). Chapter 40 Protection of the Environment; Part 141—National Primary Drinking Water Regulations; Subpart H—Filtration and Disinfection 141.72 Disinfection (a)(4)(i). Washington, D.C.: U.S. Government Printing Office.

Code of Federal Regulations (CFR). Chapter 40 Protection of the Environment; Part 141—National Primary Drinking Water Regulations; Subpart I—Control of Lead and Copper. Washington, D.C.: U.S. Government Printing Office.

Code of Federal Regulations (CFR). Chapter 40 Protection of the Environment; Part 141—National Primary Drinking Water Regulations; Subpart L—Disinfectant Residuals, Disinfection Byproducts, and Disinfection Byproduct Precursors. Washington, D.C.: U.S. Government Printing Office.

Connell, G.F. 1996. *The Chlorination/Chloramination Handbook.* Denver: AWWA.

Cornwell, D. and R. Brown. 2015. *Evaluation of Lead Sampling Strategies.* Denver: Water Research Foundation.

Cornwell, D., R. Brown, and N. McTigue. 2015. Controlling Lead and Copper Rule water quality parameters. *J. AWWA* 107:2:E86.

Deming, W.E. 1993. *The New Economics for Industry, Government, and Education.* Cambridge, MA: MIT Center for Advanced Engineering Study.

DeSantis, M.K. and M.R. Schock. 2014. Ground Truthing the Conventional Wisdom of Lead Corrosion Control Using Mineralogical Analysis. *Proceedings of the AWWA Water Quality and Technology Conference.* Denver: AWWA.

Economic and Engineering Services, Inc. (EES). 1990. *Lead Control Strategies.* Denver: AWWA Research Foundation.

Escobar, I.C. and A.A. Randall. 2001. Ozonation and distribution system biostability. *J. AWWA* 93:10:77.

Larson, J.E. and R.V. Skold. 1958. *Laboratory Studies Relating Mineral Quality of Water to Corrosion of Steel and Caste Iron (Circular 71).* Urbana: Illinois State Water Survey.

LeChevallier, M.W., O.D. Schneider, L.A. Weinrich, P.K. Jjemba, P.J. Evans, J.L. Hooper, and R.W. Chappell. 2015. *An Operational Definition of Biostability in Drinking Water.* Denver: Water Research Foundation.

Lytle, D.A. and M.R. Schock. 2005. Formation of Pb(IV) oxides in chlorinated water. *J. AWWA* 97:11:102.

Madigan, M.T and J.M Martinko. 2006. *Brock Biology of Microorganisms.* 11th ed. Upper Saddle River, NJ: Prentice Hall.

Masten, S.J., S. H. Davies, and S. P. McElmurry. 2016. Flint water crisis: What happened and why? *J. AWWA* 108:12:22–34.

McFadden, M., R. Giani, P. Kwan, and S.H. Reiber. 2011. Contributions to drinking water lead from galvanized iron corrosion scales. *J. AWWA* 103:4:76.

Nguyen, C.K., B.N. Clark, and M.A. Edwards. 2011. Role of chloride, sulfate, and alkalinity on galvanic lead corrosion. *Corrosion* 67:6:065005-1.

Nguyen, C., K. Stone, B. Clark, M. Edwards, G. Gagnon, and A. Knowles. 2010. *Impact of Chloride: Sulfate Mass Ratio Changes on Lead Leaching in Potable Water.* Denver: Water Research Foundation.

Peabody, A.W. 2001. *Peabody's Control of Pipeline Corrosion.* 2nd ed. (ed. R. L. Bianchetti). Houston: NACE International.

Rittmann. B.E. and P.L. McCarty. 2001. *Environmental Biotechnology: Principles and Applications.* New York: McGraw-Hill.

Rogers, E. (2003). *Diffusion of Innovations.* 5th ed. New York: Simon and Schuster.

Schock, M.R., A.F. Cantor, S. Triantafyllidou, M.K. DeSantis, and K.G. Scheckel. 2014. Importance of pipe deposits to Lead and Copper Rule compliance. *J. AWWA* 106:7:E336.

van der Kooij, D. 1992. Assimilable organic carbon as an indicator of bacterial regrowth. *J. AWWA* 84:2:57.

Volk, C.J. and M.W. LeChevallier. 2000. Assessing biodegradable organic matter. *J. AWWA* 92:5:64.

Washington Post. 2004. Water in DC exceeds EPA lead limit. Jan. 31.

Wheeler, D.J. and D.S. Chambers. 1992. *Understanding Statistical Process Control.* 2nd ed. Knoxville, TN: SPC Press.

Zhang, M., M.J. Semmens, D. Schuler, and R.M. Hozalski. 2002. Biostability and microbiological quality in a chloraminated distribution system. *J. AWWA* 94:9:112.

Index

For Product Safety Concerns and Information please contact our EU
representative GPSR@taylorandfrancis.com
Taylor & Francis Verlag GmbH, Kaufingerstraße 24, 80331 München, Germany

www.ingramcontent.com/pod-product-compliance
Lightning Source LLC
Chambersburg PA
CBHW070720220326
41598CB00024BA/3245